599.8
Berger, G.
Monkeys and apes.

135653

DATE DUE

	JUN 28 2001
	OCT 29 2000
	JUN 28 2000
	JUL 27 1999
	APR 15 1999
JAN 0 9 2006	DEC 22 1998
OCT 22 2005	JUN 10 1998
	APR 24 1998
AUG 24 2005	DEC 29 1997
AUG 24 2005	
JUN 11 2004	MAY 14 1997
JAN 23 2004	OCT 26 1996
NOV 07 2001	APR 06 1996
OCT 18 2001	JUN 06 1995
	DEC 22 1994

GAYLORD   PRINTED IN U.S.A.

Erich Tylinek : Photos   Gotthart Berger : Text

Monkeys and Apes

ARCO PUBLISHING, INC.,   New York

Translated from the German by Dorothy Jaeschke

Published 1985 by Arco Publishing, Inc.
215 Park Avenue South, New York, NY 10003

© 1985 by Edition Leipzig

Library of Congress Cataloging in Publication Data

Berger, Gotthart.
    Monkeys and apes.

    Translation of: Das grosse Affenbuch.
    Bibliography: p.
    Includes index.
    1. Primates. I. Title.
QL737.P9B45513 1985   599.8   84-2846
ISBN 0-668-06204-5

The chapter "Prosimians"
was written by Uta Hick

Design: Volker Küster
Drawings: Michael Lissmann

Printed in the
German Democratic Republic

Contents

Foreword

Beasts of prey, elephants and dolphins are impressive and astonishing creatures of great biological interest. But primates? Is it worth the effort of writing a big book about them? People still regard these animals, whether tree shrews, rhesus monkeys or chimpanzees, with ambivalent feelings. We admire their agility, acrobatic climbing feats and the patient care they lavish on their offspring, and laugh at their escapades or comical grimaces. We have often heard or read amazing facts about their remarkable intelligence and ability to learn. Subconsciously, comparisons are more or less inevitable and logically lead to the conclusion that man's definitely higher mental capacities and abilities are the result of an evolutionary trend which eventually developed beyond the apes to produce the human race. The anatomical features and instinctive and acquired behavior patterns of the primates reflect the phylogenic evolutionary process in which to a varying extent apes and men developed from common ancestors. This principle was recognized by Charles Darwin (1809–1882) who postulated the evolutionary theory with his book *The Origin of Species by means of Natural Selection*. This theory was also the point of departure for studies in the field of zoology, and particularly primatology, behaviorism, psychology, and anthropology. As a supporter of Darwinism, Ernst Haeckel (1834–1919), stated in a paper on *Our current knowledge about the origin of mankind* read at the Fourth International Zoologists' Congress at Cambridge in 1898: "Mankind's descent from an extinct chain of primates is no longer a vague hypothesis—it is an historical fact!"

In the course of many years' practical work with zoo animals I gained the impression during numerous talks and guided tours on the subject of "The primates up to the threshold of the human race" that even for people with a sound scientific outlook the substantiated lineage of the human race, its so-called "ape origin," is still a sensitive issue. On looking at the tupaias or some kinds of prosimians they find it hard to suppress their doubts. Fortunately, however, this attitude is definitely on the wane.

During recent decades zoo and laboratory research and also extensive and systematic field studies have greatly enriched our knowledge about the biology, behavior and abilities of the primates. It is only necessary to pick up *Brehms Tierleben* of 1916 with its classical description of the primates or the standard work by R.M. and A. Yerkes *The Great Apes* (1929) to realize this. New and sensational progress, especially in behavioral research, has been made and recorded in detail by Jane van Lawick-Goodall, I.S. Bernstein, I. De Vore, R.A. Hinde, A. Jolly, M. Kawai, A. Kortlandt, M. Moynihan, V. Reynolds, T.E. Rowell, G.B. Schaller, Y. Sugiyama, S. Ripley, B. Rensch, I.T. Sanderson, and S.L. Washburn whose names stand for those of many other scientists.

As far as Europeans are concerned, their opportunities for making the acquaintance of our nearest relatives in the animal kingdom and for learning more about them is generally restricted to the larger zoos and wild life parks that contain many primate species. That, too, helps to remove prejudices.

But the primates are not only biologically interesting; they are also endearing creatures in every respect. They are worthy subjects for a book of this kind.

In compiling this book I endeavored to include as much material as possible and considerable abridgements of the completed manuscript were therefore inevitable. I hence make no claim to have covered every aspect of the subject. It is primarily a survey of the wide variety contained in the order of primates with some more detailed examples that can be read separately by those seeking information about the species concerned. The bibliography gives a list of specialized works for readers requiring in-depth knowledge about primatology.

With a few alterations I based the classification on the one published in *Grzimeks Tierleben*, Vols. X and XI, Kindler Verlag, Zurich, 1968.

Introduction

In recent decades primatology has become one of the most important and fascinating fields of zoology. Not only because, as the name indicates, it concerns a group of the most highly developed animals, but also because it includes mankind too. As early as 1758 the Swedish naturalist Carl von Linné (Linnaeus), the founder of scientific zoological and botanical classification, admitted the feasibility of a certain degree of kinship between primates and men. But a clear and rational explanation was at that time not possible. In 1863 T. H. Huxley stated: "Perhaps no order of mammals presents us with so extraordinary a series of gradations as this—leading us insensibly from the crown and summit of the animal creation down to creatures from which there is but a step, as it seems, to the lowest, smallest and least intelligent of the placental mammals." Indeed, this order of mammals comprises a far greater variety of forms than is generally associated with monkeys, ranging from the ancestral shrew-like types no bigger than a rat to the gorilla that weighs up to 250 kg and the chimpanzee with its close resemblance to human beings. It is therefore no exaggeration to state that a review of all existing primate species provides a cross-section of their entire lineage. "Evolution," stated the genetic expert T. Dobzhansky, "is a combination of determinism and chance, and this combination makes it a creative act . . ." Up to the present day the primates have demonstrated in many flexible variations their powers of survival and have produced the most highly developed living beings. Their mutations and variations provide convincing examples of evolutionary dynamics as a driving force within the order of primates and help us to establish phylogenic links. The biological study of primates is therefore hardly feasible without a side-glance towards mankind. Today there are about 190 distinct primate species classified into 15 still existent families.

Linnaeus' brief definition of the primates is still substantially valid: "Four parallel upper incisors, an eyetooth on each side, two mammary glands on the chest, limbs suitable for gripping, a collar-bone forming an important supporting element for the function of the arms, movement on all fours, the climbing of trees and gathering of their fruits." De Vore put it more bluntly: ". . . there is in fact no fundamental criterion by means of which the whole group can be defined . . . The only aspect that can safely be regarded as a common feature of all living and extinct primates is their adaptation to life in the trees." To this can be added the gradual process that began with the prosimians of the transition of the eyes from the side of the head to the front. This effected a material improvement of binocular-stereoscopic sight; eye-sockets completely encircled by bone; the improvement of sight accompanied by a loss of the sense of smell; the cerebrum is deeply furrowed; at least one pair of limbs has a first finger or toe (thumb or big toe) that can be opposed to the other digits; fingers as well as toes generally have nails instead of claws; a well-developed caecum; the testes descend into a scrotal sac; the relaxed penis usually droops; highly developed brain capacity up to the threshold of thought; a communal life-style with a more or less pronounced order of social precedence; outstanding ability to defend themselves on a group basis; vegetarians or omnivores.

At least 30 million years ago geological processes effected a very significant division, namely into New World primates (South and Central America) and Old World primates (Africa and Asia). The primates adapted themselves to the most varied environmental conditions (in terms of nutrition, strength and defensive powers, ecology) and were thus able to survive.

Men and Primates

Early ideas and monkey worship

Animals were the first subjects that human hands ever painted on rock faces and the walls of caves—not just because men hunted them and they were a source of food but primarily because they were alien and fascinating creatures that possessed better physical qualities than men in terms of strength and speed—creatures that could swim and even fly. Animals that inspired such admiration rose to the rank of gods or their representatives. Our ancestors regarded the monkeys' resemblance to humans not as a matter of chance but as a kind of divine providence, so that certain species found a place in their mythology or were worshipped in various religions.

These original interpretations are chiefly known to us from the early civilizations and to a lesser extent from the primates' ranges—the extensive forests of the African and South American interior. Religion in Ancient Egypt that assigned a soul to all living beings and whose gods appeared in animal form, portrayed, for instance, the hamadryas baboon on countless murals and sculptured it sitting upright. As a sacred animal, the servant and companion of the god Thoth, it gave men understanding and knowledge and was also the spokesman of the creator of all things. It was portrayed as the god of scribes too, and although not a human being was regarded as a priest (see description of hamadryas baboon). A stone statue from the 19th Dynasty (ca. 1350–1200 B.C.) depicts two baboons worshipping the sun. The statue of the goddess Toeris from Karnak (Upper Egypt), ca. 500 B.C., resembles a baboon standing upright. A sacred baboon depicted on a papyrus symbolizes the link between the great eye of the god and his human worshippers. In Egypt, magnificent temples, groves and lakes were consecrated to the god-animals embodied by monkeys, ibises, serpents, cats and other creatures. Near Thebes there was a monkey tomb. It is not known whether there is any connection between "hamadryas," the name given to this baboon, and the hamadryads, the graceful tree nymphs of the Ancient Greeks.

In its portrayal of satyrs early Greek art underscores the grotesque simian elements. It was not until later that the monkey-like satyr became a dreaming youth. The Greek historian Diodorus (1st century B.C.) described, according to H. Wendt (1956), the Egyptians' monkey cult as follows: "Pan and the satyrs were worshipped for their progenitive powers. By portraying them people offered up thanks for the blessing of progeny. . ." Diodorus also gives a well-observed description of the African baboon: "The so-called dog's head makes the creature resemble an ugly human being and its voice, too, sounds like human groans. These animals are extremely savage and quite untameable, and their eyebrows give them a sinister appearance. The females have a very remarkable feature—their wombs are always outside their bodies." This observation probably refers to the typical swellings of the sexually receptive female baboon that often look like malignant tumors. In medieval times the Egyptian Sphinx, half-human and half beast of prey, was thought to have a monkey's body. This interpretation of its shape was possibly the result of confused reports about monkeys. Even less adequately explained are the puzzling accounts of the Gorgons, legendary creatures of the Greek Classical Age with supernatural powers whose awesome heads can be found on many pictures of those days. According to Ancient Greek sources Libya was the original home of these distorted humanoid-animal heads with huge dagger-like eye-teeth and protruding tongues. This western region adjoining North Africa was under Carthaginian-Phoenician influence. Hanno, a Carthaginian admiral, and his fleet reached the West African coast (in the region of today's Gabon) in the early 5th century B.C. According to his account they met a savage forest tribe whom the interpreter called gorillas. So possibly the Gorgon depicted on a Grecian vase represents an African ape. Another Gorgon can be seen on the handle of an Attic amphora dated about 540 B.C. We know of a much older piece of evidence; a silver bowl found in an early 7th century B.C. Etruscan tomb in Praeneste which belongs to a whole set of Phoenician or Carthaginian origin. Its outer frieze depicts a number of hunting scenes, one of which distinctly shows an upright and very hairy figure. It holds a stick in one hand and in the other a stone which it is about to hurl with a typical arm movement; shape, stance and appearance are identical with those of today's chimpanzees and gorillas (see A. Kortlandt 1967). Typical features here are the body proportions corresponding to those of a great ape (gorilla?). "Today this silver bowl is considered to be one of the earliest pieces of evidence of civilized man's encounter with his next relatives in the animal kingdom" (H. Wendt 1971). In the rock paintings in the South Algerian Ahaggar Mountains a picture dating from the 4th millennium B.C. is thought to depict a "gorilla-god." Its shape actually resembles that of a gorilla. Very probably it was worshipped at this place. In later times, too, the huge proportions of the gorilla evoked the respect and homage of African tribes. The pygmies, for instance, had similar concepts. If gorilla babies fell into their hands while hunting, these were sometimes breast-fed and reared by the women. Adult male gorillas are five times as heavy and twice as large as the pygmy hunters who worship them. Even today isolated tribes wear ape masks for certain cult rituals in order to exorcize demons, forest spirits, etc.

Previously the natives of Madagascar regarded certain prosimian species as sacred animals. When basking in the morning sun the ruffed lemurs as well as the ring-tailed lemurs resemble sun-worshippers—they sit upright, stretching out their legs and turning their faces toward the sun. The indris and sifakas do the

Cynocephalus—human being with a dog's head in a natural history book dated 1642.

same. The equation of apes and men is often based on the very distinct resemblance to human beings. The Malays, for instance, do not consider the big red ape to be an animal but a "forest man," in the Malayan language "orang utan."

"Orang pendek" means a pygmy, a general designation for the aboriginal tribes who were short-statured; in contrast, the Malays who arrived later named themselves "orang melaju"—"migrating people"—to distinguish their own culture and origin. Various Indio legends cited by Alexander von Humboldt tell of great man-like apes. Two stone statues from Maya times in the Museum of Archaeology and History in Merida (Mexican Province of Yucatán) show a surprising resemblance to a gorilla and chimpanzee. There is also a photograph taken by the Swiss geologist François de Loys on the banks of the Río Catatumbo in the tropical forest of Tarra (Venezuela) between 1917 and 1920 which shows another mysterious primate. The animal is said to have had reddish fur and to have more closely resembled a human being than any other Venezuelan monkey. Its height remained controversial. Could it have been a large specimen, that had somehow lost its tail, of the red howler monkeys that live there? The question was much discussed but the existence of this mysterious man-like ape named Ameranthropoid was never corroborated. Even later reports about other "ape-men" of this or that kind did nothing to make these statements more credible. At the turn of the century there were also some very single-track zoologists, like the Argentinian F. Ameghino, the Dutchman Bierens de Haan and others, who, taking as their point of departure the intelligence displayed by the capuchin monkeys and some very dubious bone remains of "early man-like apes," tried to transfer the cradle of humanity to South America.

At the beginning of modern times the first reports about man-like apes began to reach Europe, and tales of hirsute animal-men rapidly spread. In pictures they were given human proportions but were often still shown with a long tail. Around 1590 the Englishman A. Battel described his encounter with great apes on the Loango coast in the region of today's People's Republic of Congo: "The forests are so crowded with baboons [probably drills and mandrills—the author], guenons, monkeys and parrots that everyone fears to travel through them. There are in particular two sorts of monsters frequently found in these forests that are extremely dangerous. The larger of these horrors is named pongo by the natives, the smaller ensego." "Pongo" definitely refers to M'pungu, the gorilla, while "ensego"—later on in the 19th century called "chego"—is the chimpanzee. The first living chimpanzee came to Europe in 1641 in the menagerie of the Prince of Orange and then became the object of scientific studies. Even after the Middle Ages, mythology, reports about exotic tribes and descriptions of a host of ape-like figures were so mixed up that it was hard to find a grain of truth. In the age of discoveries, explorers, seafarers and traders brought more authentic reports to Europe. The Portuguese envoy P. Alvarez, for instance, presented some very accurate observations and descriptions of large baboon groups in Ethiopia.

The best known monkey in Classical Antiquity and the Middle Ages was the magot (Barbary ape). K. Gesner was the first to give a fairly accurate description of it. Seafarers brought the "markata," the Indian rhesus monkey, to Europe.

Reverting to the "Darkest Africa" approach, P. Du Chaillu, an American hunter and explorer, described the gorilla in 1855/56 as a beast possessed by demons. W. Reade and later C. Akeley, M. Johnson and J. von Oertzen corrected this horror story.

The mythology of South and East Asia contains a multitude of demons in half-human, half-monkey guise. For instance, in many legends the king of the monkeys comes to the rescue in battles. Hanuman, the monkey prince, is a figure that often adorns the handles of Burmese daggers and the Indonesian kris. Together with other animals many monkeys can be seen on the frescoes in the cave temples of Ajanta. The monkey princes and kings were considered to possess great magical powers and the ability to transform themselves at will. Moreover, Hanuman was thought to be able to hurl himself across vast distances. The leaps performed by Hanuman and the other monkey princes in the *Ramayana* are a mythical symbol, the equivalent of Garuda's flight, for in actual fact these leaps are really flying. Thus Hanuman leaps a hundred miles away from the coast across the sea to Lanka and from there to the Himalayas and back. In the Oriental shadow plays, an art which held a deep fascination for the Indian and Chinese culture, one of the puppets is Hanuman, the shrewd minister of the monkey-god Sugriva who was the comrade-in-arms and friend of Rama. These Javanese wajang figures of Hanuman and the monkey king have a formidable appearance with ferocious-looking jaws. In various guises Hanuman is also to be seen in Balinese sculptures. In the Vat Benchamabophit temple in Bangkok there is a cabinet inlaid with mother-of-pearl with figures portraying the élite of the monkey dynasty headed by Sukhrib, the king, Hanuman, Ongkot and Nilaphat (early 18th century). Devout Hindus still worship the hanuman monkeys, although not to the extent they did in former times, as the incarnation of the monkey-god Hanuman (see description of the hanuman monkey). But in their range, the rhesus, bonnet, and hanuman monkeys have almost become a "holy plague." In South and Southeast Asia close links between men and monkeys have existed for a very long time and are to a certain extent reflected in these peoples' cultures. In many Hindu homes we find amongst the colored portraits of divinities that of Hanuman, the monkey-god who is also the patron of the magic and medical arts. The monkeys are chiefly found near the temples and places that attract large numbers of tourists where the animals can be sure of getting food. They sit on the temple statues and balustrades, scamper in troops through the streets, explore every nook and cranny outside the doors and shops and, when given a chance, steal their favorite foods and other articles. Their status as a sacred animal prohibits any kind of forcible restraint or punitive action. Sometimes macaques are caught and taken to the forests. Frequently, however, they return to their old homes. Where they enjoy special protection, such as at the Benares temples on the Ganges, at Adjodhya in northern India, the holy city of the legendary King Rama or other crowded places of pilgrimage, the monkeys are often so bold and impudent that shops, washing hung out to dry, open vehicles, etc. have to be protected from their inroads and attacks. Monkey cult solo and mass dances are performed in Indonesia too, for the benefit of tourists as well as believers, for instance the Hindu monkey dance in honor of Hanuman on Bali.

Not only the hanuman and rhesus monkeys, but also other macaques, such as the crab-eating macaque, enjoy immunity as sacred animals. Japanese macaques are considered to be a symbolic incarnation of the wisdom of life in the Buddhist religion. In the very widespread fox mythology in China and Japan, the monkey Songoku, which bears a close resemblance to the Japanese macaque, plays an important part with his magical powers. The Japanese macaque is portrayed as an object of veneration in numerous pictures and woodcuts in its homeland.

Another magot achieved a kind of symbolic significance in the British Empire, best expressed in the traditional saying that the British will only lose Gibraltar when the last Barbary ape there has disappeared. Today the Gibraltar monkeys are still under the jurisdiction of the British military authorities. It is quite possible that the Phoenicians, Carthaginians or Romans brought the first magots to Gibraltar, letting them run wild on the "European pillar of Hercules."

Both in the mythology and art of China animals have played an important role from the earliest days. Monkeys, considered fabulous creatures for a long time, were portrayed on vases dating as far back as 2200 B.C. and also later when the living animals had been discovered. An example is the snub-nosed golden monkey. In describing this species hitherto unknown to zoology, A. Milne-Edwards was reminded of a 16th century portrait of the Russian courtesan Roxellane. Because the lady had auburn hair the color of the monkey's fur and a charming snub nose, the zoologist's sense of humor inspired him to name this Chinese monkey *Rhinopithecus roxellanae*. Was a Chinese painting from the time of the Ming Dynasty (15th century) the product of rumors or was it based on more or less reliable eyewitness accounts? It depicts a half-human, half-animal creature with a dark, thick, long-haired coat. Looking like a "furry demon" it recalls the ferocious flat-nosed figures on a Buddhist or Tibetan temple. Or is it an early attempt to portray the "abominable snow man," the Yeti, who is said to live in the higher regions of the Himalayas? Since that time there have

been many reports, photos and drawings of footprints in the snow, descriptions, speculations and even encounters. From East Tibet via Bhutan, Sikkim, Nepal, up to the Karakorum Mountains, the rumors obstinately persist, nourished by the often grotesque accounts of the indigenous population. After careful examination, however, the footprints turned out to be made by bears, mountain hanumans, snow leopards, hermit lamas, etc. On the other hand there are fairly credible reports made by explorers and mountaineers. In 1925 A.N. Tombazi, an Italian mountaineer who was attempting to ascend the Zenni glacier in northern Sikkim, had his attention drawn by his coolies to a dark, naked, man-like figure who was wandering upright along the valley between the rhododendron bushes. "I examined the footprint and found it resembled that of a man but only measured 18–20 centimeters*. I could clearly see the imprint of five toes and the sole." More or less similar reports were made in 1936 by the ethnologist and botanist R. Kaulback. But zoologists, too, gradually began to take an interest in the mysterious snow man. After 1945 other impartial witnesses reported what they had seen: E. Shipton, head of the British Mount Everest Expedition of 1951/52, accompanied by Hillary, the later conqueror of Mt. Everest, and the celebrated Sherpa bearer Sen Tenzing, had discovered strange footprints in a valley 6,000 meters up on the Everest massif. Tenzing immediately declared that a Yeti had made them. Right up to the present day this kind of report persists. In 1974 a news item from Katmandu stated: "The Nepalese police are making progress in their search for the Yeti . . . who at the foot of Mount Everest is said to have struck down a 19-year-old Sherpa girl and killed five yaks . . ." The Soviet scientist Pronin, who years ago encountered a Yeti in the Pamir uplands, even managed to make a sketch which depicted a creature looking like a very hairy Neanderthal man. Unfortunately such reports are often sensationally exaggerated in the press and coupled with speculations of an unscientific nature. Another report early in 1982 stated that an Indian military patrol in the Chetak Pass frontier region bordering on Bhutan had discovered extremely primitive cave-men who were ignorant of clothing and how to make fire. They ate their food raw. A solution to this riddle will probably be found before long and perhaps it will also provide an explanation for the mysterious blackish-red scalp in the temple at Pangbotchi. So even in our days the animal kingdom still presents some unsolved mysteries.

---

* For conversion factors see page 240.

Primates in the service of mankind

Even in Ancient Egypt people recognized that primates are useful animals and turned their abilities to good account. Baboons were the first "harvest hands." They helped humans to pick fruit, for instance coconuts. Their agility in clambering about in the trees, in grasping objects, their ability to learn easily and to distinguish between ripe and unripe fruit predestined them for this activity. In Southeast Asia, using the method of rewards for good results, the pig-tailed macaque has been successfully trained to pick coconuts. A hard-working macaque can manage to pick a thousand nuts a day. The owner of coconut palms either has his own macaque, hires one from a neighbor or from a professional trainer who makes his living in this way. The owner of a macaque treats it as one of the family. Incidentally, the coconut, which chiefly grows in the countries bordering on the Pacific, owes its name to the "macocos" (macaque; see description of pig-tailed macaques). In their natural habitats baboons are today used for harvesting purposes. In Thailand, Malaysia and other Indo-Pacific countries, these animals are also useful in field studies connected with the forest flora. They are trained to throw down specimens of all the plants growing high up in the trees. In Southwest Africa chacma baboons have been successfully taught to act as goatherds (see page 133). Attempts to train chimpanzees, the most intelligent apes, to perform similar useful work, have failed.

The flesh of monkeys was indubitably eaten by human beings very early on. Indio tribes today still hunt monkeys as a source of food. These are mostly capuchins or wooly monkeys that are killed by poisoned arrows or trapped. On the other hand, they are also tamed and kept as domestic animals and pets for the children—especially marmosets and tamarins. In those regions of South and Southeast Asia where monkeys have no religious or other cult significance, their flesh is also eaten. This practice has brought, for instance, the already depleted number of lion-tailed monkeys and John's langurs in South India to the verge of extinction. In Africa, too, monkeys are hunted and eaten. The fur of these animals was and still is utilized—although today to a lesser extent as a result of strict protective measures. Amongst the Madagascar lemurs those with beautifully colored markings, such as the ring-tailed and ruffed lemurs, were much coveted by fur traders. Because of the difficulties of catching these animals and their small numbers only relatively few furs were obtained. The same thing happened to the slow loris that lives in South and Southeast Asia. Stringent protective measures on Madagascar put an end to this destructive practice there. About 5,000 hides annually used to be delivered to the fur trade. In South America the fur of the black howler monkey is often used for caps, blankets and bags. The magnificent long-haired black and white fur of the guereza, particularly the mountain varieties, almost led to the extinction of these animals. Around the turn of the century guerezas were sacrificed to the dictates of fashion. Almost a million of them were killed before and after the outbreak of the First World War in consequence of a fashion launched by the international trendsetters in Paris, London, and New York. Monkey-fur coats, jackets, waist-coats, capes, muffs, etc. were all the rage. Around 1970 a comeback of this fashion launched from Paris was frustrated thanks to the better organized protection of these animals. Nonetheless, as a result of the growth of tourism, there is still a relatively big demand for these furs (see description of the Abyssinian guereza). After the Second World War there was a market principally in the Western countries for the smaller South American species like the squirrel monkey, marmoset, and tamarin. Unfortunately social prestige rather than biological interest was the main incentive for the acquisition of these unusual pets. Many of these delicate monkeys perished because they were not kept under proper conditions. In Baroque times the possession of a monkey was regarded as a status symbol amongst the aristocracy and other wealthy circles.

Similar to the use of guide-dogs (Alsatians) for the blind, attempts have been made to train suitable monkeys to help severely handicapped people. For instance, a specially trained young adult capuchin (1980/81) successfully helps a young man in Massachusetts whose limbs are completely paralyzed. The animal can put the patient's food in a microwave oven to warm, take bottles out of the refrigerator, open them and put a straw in the man's mouth, comb his hair and perform other simple services. It naturally takes a great deal of time and effort to train a mobile and easily distracted animal like a monkey to do such work reliably, and it is unlikely to become general practice. As long as alternative facilities are available, the chief value of such experiments will continue to lie in the sphere of basic research.

Primates in biological and medical research

Primates today occupy a very important place in biological and medical research, particularly in behavioral studies (ethology), physiology, psychology and anthropology. One aspect of these studies is the drawing of philosophical conclusions from the

---

1 Crab-eating Macaques (*Macaca irus*) at the temple of the sacred monkeys in Surabaya on Java. Because they enjoy complete protection these monkeys sometimes pester human beings.

2 "The Battle with the King of the White Monkeys," a symbolic dance in Sanur on Bali.

3 In the National Park at Nikko—about 150 kilometers from Tokyo—is the Stable of the Sacred Horse. Among the mural carvings by Kano Tany-yu (1608–1674) the representation of the three monkeys that hear, speak and see no evil is known the world over.

findings of modern biology in general, and primatology in particular, and the clarification of ethical problems in this context. "These are problems of how to impart a dialectical approach to biological thought in view of the progressive discovery through biological studies [here in primatology—the author] of the dialectics of living beings," as R. Löther stated (1980). Because it is a question of the humanist-ethical orientation of research methods, there is a moral obligation involving responsibility and respect for life. Experiments on living primates in the sphere of medicine and the pharmaceutical industry as well as in space research must comply with this stipulation, and the same applies to the specific biological needs of the animals kept in primate centres and breeding farms that supply "test material." Many primate species ranging from the tupaias (tree shrews), lemurs, marmosets, guenons, capuchins, and vervets as well as chimpanzees are today used as laboratory animals. "Primate-tested" is regarded as a hallmark for the quality, safety and reliability of a medical drug.

There is convincing evidence that experiments with living primates have indeed served the welfare of humanity. But in recent times demands have been more frequently expressed that such experiments should be far more restricted or entirely abandoned. These demands are principally motivated by moral considerations—the responsibility of the human race for life in general.

Animals which have more highly developed brains not only experience pain more intensely but are also acutely affected by psychological factors such as anxiety, stress, intimidation, agitation, fright, shock, etc., which involve lasting torments for these test animals. This kind of stress is equivalent to performing life-endangering operations on and in the body of the animal, which is a helpless victim since it cannot follow its instinctive urge to flee when frightened. There is ample evidence demonstrating that in certain investigations significant results can be achieved with the use of modern biotechnological or biochemical methods, so that animals are not essential for such experiments.

Let us take a closer look at some medical and biological research results. Thousands of young people would still die of the dreaded poliomyelitis if a vaccine had not been discovered whose effects were first tested on chimpanzees.

Like every other virus the polio virus only grows in living cells. Fresh tissue therefore is needed for the cultures, particularly the kidneys of recently killed rhesus monkeys. With the introduction of oral inoculations this terrible disease has been practically eliminated.

---

4 Our zoos, too, are not infrequently embellished with handsome sculptures of monkeys. (Prague Zoological Gardens)

The Rh-factor, which is inherited irrespective of other blood factors, was first discovered in laboratory experiments on rhesus monkeys after which it was named. As a result of this discovery specific damage to the human embryo can today be prevented. Biological and medical research, for example in the Primate Laboratory of Oregon, is concentrated on finding ways to cure the most dangerous diseases that beset the human race. Some of the areas of research are muscle dystrophy, skin diseases, the effects of hormones during and after pregnancy, cardiac diseases, brain electricity, arteriosclerosis, tissue transplantations and the connections between viruses and cancer. All the daily control data from far more than a thousand monkeys are processed by a computer. An unusual experiment with rhesus monkeys has helped to explain the heavy toll of casualties in today's civilization whose stress and pressure lead to neurotic disorders. A remarkably large number of people die at a relatively early age when they have been subjected to great stress over a longer period. It is thought that this is a result of the insufficient elimination and consequent accumulation of adrenalin which saturates the blood under stress conditions and if physical exercise is lacking as a compensatory factor. When primates have been subjected to temporary stress they work it off through physical exercise. The highly virulent and up to that time unknown germ which caused the "Marburg monkey epidemic" in 1967 when five of 22 infected people died, was found to have been imported with vervet monkeys from Uganda. Well-known research centers in Europe and America made a special study of this epidemic particularly after 1976 when once again this highly infectious and very mutable virus disease took a heavy toll of lives in Sudan and Zaire. It was discovered that monkey tissue produced intense infection amongst guinea pigs.

In Cape Town in 1978 Dr. Barnard transplanted a chimpanzee heart for the first time into the body of a cardiac patient, who however died. Nevertheless, intense research is still being carried on in a number of other countries too with a view to using primate organs as transplants for the human body. One of the largest and most celebrated of the many regional primate research centers in the United States, including those in Seattle, Southborough and Beaverton, is the Yerkes Regional Primate Center—named after R. M. Yerkes (1876–1956), the famous primatologist—attached to the Emary University in Atlanta. In 1973 more than 140 chimpanzees, orang utans, and gorillas were kept for biomedical research purposes. In the sphere of psychology as well, extensive studies and experiments over recent years have produced evidence about the intelligence, personality development, and social behavior, particularly of apes. Founded in 1927 and today known the world over, the Sukhumi Primate Center of the Institute for Experimental Pathology and Therapy of the Academy of Sciences of the USSR has in recent years (1979/81) kept about 2,000 primates belonging to 15 species, including many baboons, on an area covering 126 hectares. Large parts of the Primate Center itself are well known as a public popular scientific and biological educational establishment. Here the animals are chiefly used for research purposes concerned with cancer, leukaemia, cardiac and circulatory diseases as well as for organ transplantations. Evidence has been produced, for example, to corroborate the theory that cancer of the blood, etc. may be caused by viruses. Intensive efforts, with international participation, are also being made to find an anti-cancer vaccine. Behavioral studies are carried on in Sukhumi too. Acclimatization experiments (including frost and snow) were carried out on about 70 hamadryas baboons 350 kilometers north of Sukhumi, in a forest on the southwestern slopes bordering the Black Sea. The ability of these primates to adapt themselves was demonstrated by their much thicker and longer fur as compared with their peers in Sukhumi and a relatively large number of progeny.

H. Harlow at the Wisconsin Regional Primate Research Center made specialized ethological and physiological studies of primates' mental activities, reactions, and how these are signalized, with a view to establishing their resemblance to human emotions. This research on chiefly young rhesus monkeys has helped to answer the question of how far primates are influenced in their behavior by feelings, affection, emotions and love (mother love) and, first and foremost, whether they possess these qualities at all. The program included extensive systematic research into the ability and talent for speech, the aptitude to learn, memorize and reason ("thought without speech?"), to abstract and act rationally, social structures, aggressive behavior, use of tools, rearing of the young, training

results, and laboratory conditions, adaptation processes, etc. At Göttingen University, the Institute of Anthropology under Professor C. Vogel is making a study of groups of marmosets and tamarins with special emphasis on reproductive and biological aspects in connection with the social order. Considering the close relationship between apes and humans, can these species be crossed? Responsibly-minded geneticists and physicians reject such experiments on ethical grounds. Such a hybrid, even if it were able to live, could never develop normally, which is why this kind of experiment with the aim of creating a living foetus, let alone a birth, can hardly be justified. The comparison of human and anthropoid sperm offers the opportunity of casting more light on the phylogenic kinship between man, gibbon, chimpanzee, gorilla, and orang utan. The comparison of the morphology and DNA content of sperm from the above-mentioned species undertaken at the Western General Hospital in Edinburgh produced not only similarities but also marked differences.

Primates in the conquest of space

Humanity's first steps towards outer space were accompanied by new areas of biomedical research. Before human beings could risk the hazards of space flight an "understudy" had to be found on which all the hitherto unknown effects of a weightless state, radiation, acceleration pressure, etc. could be tested. Besides mice and guinea pigs the Soviet Union used dogs as cosmonauts, e.g. the celebrated "Laika" in Sputnik 2 in 1957. In 1949 a rhesus monkey survived a flight at a height of 134 kilo-

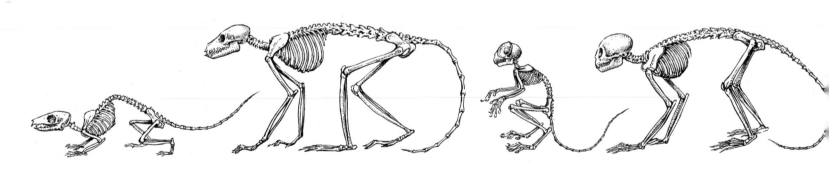

Tree shrew   Black lemur   Squirrel monkey   Tarsier

meters in the capsule of a US ballistic rocket. In 1959 it was followed by "Miss Baker," a spider monkey, in the nose of a Jupiter rocket that made a similar ballistic flight into space. Despite her hazardous astronautic feat—she has lived since then in the Huntsville Rocketry and Space Flight Museum—she reached the age of 25 in 1981 (average age 20 years). The rhesus monkey "Lizzie" paved the way for further biological preparations for the first US space flight. Kept for 90 days in a chamber filled with pure oxygen she had to carry out numerous physical exercises and switching tests. The purpose of this experiment was to find a suitable atmosphere that would enable the future cosmonauts to breathe freely. "Ham," a four-year-old chimpanzee and the first NASA "chimpanaut," was launched from Cape Kennedy on January 31, 1961 in a Mercury capsule traveling at a speed of 8,000 kilometers per hour and reaching a height of 248 kilometers; this rocket, however, did not go into orbit, the flight lasting only 16 minutes. "Ham" was fitted out with a pressure suit and many biomedical control devices. Now also 25 years old (1982), "Ham" lives in the Washington National Zoo. Remarkably teachable and quick on the uptake, "Ham" naturally underwent a very complex and strenuous training program calling for high standards of physical and mental performance at the Aeromedical Research Laboratory of the Holloman Airforce Base in New Mexico before his flight. Doubts were already expressed at the time as to whether such abnormal demands were justifiable, but apart from the actual aim of the program, an astoundingly high level of intelligence was noted, such as, for example, the mastering and logical operation of control systems with levers and press-buttons in response to commands and a sequence of flashing signs on a panel. In No-

The skeletons of various primates illustrate the different postures and sizes (after Jenkins, Hill, Schultz, and others).

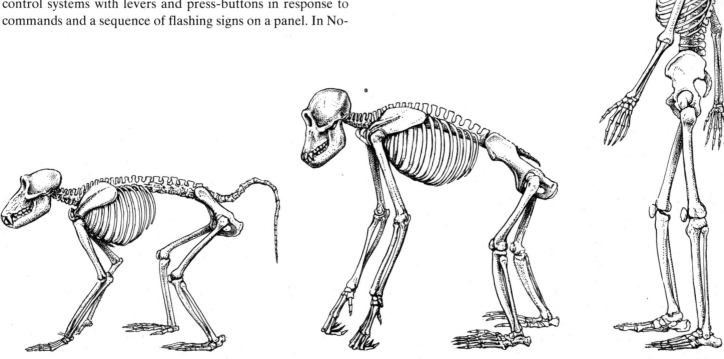

Baboon   Chimpanzee   Man

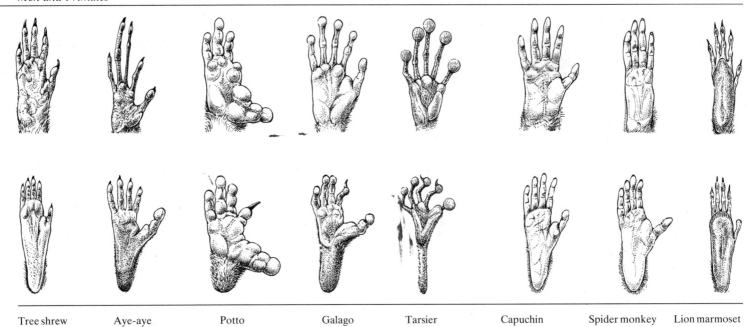

| Tree shrew | Aye-aye | Potto | Galago | Tarsier | Capuchin | Spider monkey | Lion marmoset |

Hands (top row) and feet (bottom row)
of various primates,
all reduced to the same length (after Schultz).

vember 1961, "Enos," another rhesus monkey, was launched into space as a pioneer for mankind.

The use of primates for military purposes is an abuse that must be condemned on principle quite apart from the fact that the animals are harmed to an extent that usually precludes survival (e.g. radiobiology tests, hot and cold shocks, being kept in low-pressure chambers until a vacuum occurs, etc.).

Some examples taken from "Ausverkauf der letzten Affen" (Selling off the last primates), an article by H. Hediger that appeared in the periodical *Das Tier*, No. 7/1976, help to illustrate the "mass consumption" of these animals. The statistics contain some discrepancies, but in 1958 the USA alone imported 223,000 primates (W. Starck 1959). These, however, are just the officially registered transports. Via black market channels many thousands of primates reached the recipient countries without registration. As a result of frequently inadequate transportation in very confined cages many animals die before reaching the recipient and at the often overcrowded reception centers and depots a large number of animals perish because of exhaustion, bad hygienic conditions and parasitic diseases.

Moreover, in the case of the larger primates, such as apes and baboons that are capable of defending themselves, the parents have to be shot before their young can be captured. In 1969 Peruvian officials estimated 30 to 50 percent losses in the capture, acclimatization and transportation of primates (B. Harrison 1976). The Pasteur Institute in Chile estimates that for every exported healthy young chimpanzee between four and six mothers are killed. Similar figures for East Africa were estab-

lished by J. van Lawick-Goodall. When 125 chimpanzees were ordered from Sierra Leone for liver research purposes it was stipulated that a hundred of them were to be females at least one year old. So about double this figure had to be captured since it was estimated that approximately 50 percent would be male children. Hence for this project at least 750 chimpanzee mothers had to be shot! In 1978 alone 1,342 imported crab-eating macaques died after great suffering at a research center in order to establish the actual toxic properties of a herbicide that had previously been tried out on dogs and rats. W. Conway, general director of the New York Zoological Society, uttered a warning many years ago about the effects of the wanton depletion of the primate herds in South American, African, South and Southeast Asian countries. According to Conway (1966) there were already less primate imports into the USA in that same year. Not reduced demand but delivery difficulties, limitation of export permits as well as rising prices were the reasons. Up to the nineteen-sixties India was the chief exporter of rhesus monkeys particularly for the production of anti-polio vaccine. In 1960/61 about 250,000 rhesus monkeys were exported annually, in 1963 only 150,000 and in 1964 a mere 11,000, a figure to which bonnet monkeys must be added. Bangladesh, a very poor country, later offered to export 3,000 rhesus monkeys annually. Since these monkeys had become a world-wide "scarcity"—and not just for laboratory purposes—the rational solution (but not necessarily a biologically wise one, as Professor Starck commented) was to use other species. Hence one East African country exported about 25,000 guenons in 1962

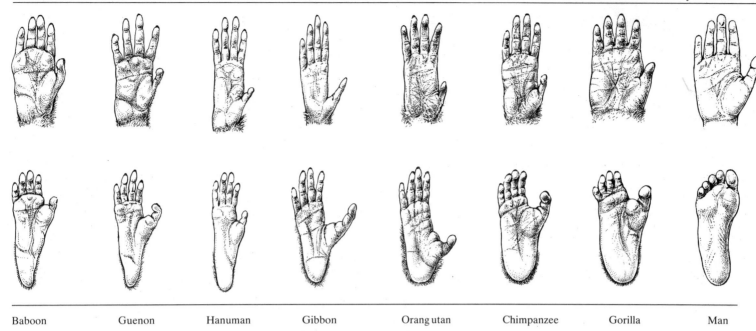

| Baboon | Guenon | Hanuman | Gibbon | Orang utan | Chimpanzee | Gorilla | Man |

and in the late nineteen-sixties 18,000 vervets. Other countries followed suit to a lesser degree. Even the rare owl-faced monkey was not spared, and South America supplied increased numbers of squirrel monkeys and moustached tamarins. There was a growing demand for more macaques, pig-tailed, and crab-eating macaques from Southeast Asia too. Since then the main recipient countries have naturally attempted to accelerate the breeding of primates at their own centers. In countries with a temperate climate the establishment of proper biological conditions involves much outlay in terms of building, heating, and zoological factors. Macaques and guenons do not reproduce until they are five or six years old and only give birth to one young at intervals of at least two years. Apes are even harder to breed in captivity. According to a report given by H. Hediger (1976) there were 164 chimpanzee births at Yerkes Primate Center between 1930 and 1966. Although India, for religious reasons too, stopped or much reduced the export of rhesus monkeys particularly to the USA which misused them for military purposes (1978), this profitable trade was continued via third countries. In 1976 it was planned to establish a monkey breeding farm simulating natural conditions near Bombay in order to increase exports without further decimating the already depleted indigenous primate herds. In some European cities very extensive primate breeding centers have been built in recent years. The aim is to rear many hundreds of primates, mostly rhesus monkeys, annually. For economy's sake, many of the animals are housed separately in very small cages. Close confinement and the inability to move around is probably the cruellest kind of treatment for such a mobile and sociable animal. Most of the primates bred in such centers are intended for the pharmaceutical industry. There are, however, other laboratory breeding centers that provide better conditions for the animals. The countries that previously had the highest requirements now import about 20,000 primates annually—approximately eight percent of the 1960 figure. It is to be hoped that in future primates will only be used to meet the most urgent laboratory needs and that before long this practice can be completely abandoned. "Let us beware of irresponsible behavior with the lives of primates. We are the only species of this order that can do anything for the survival of the other species. After all, we ourselves have given our species the name of *Homo sapiens*. We should call this to mind now and then." (H. E. Schneider 1979)

There is yet another way in which human beings make monkeys and apes work for them. In bygone times and even today, particularly in South Asian countries, exhibitors and beggars earn a living by putting trained monkeys, mostly macaques, through their paces on busy streets and markets. Another alarming abuse is found in the tourist centers on the Spanish coast and on Tenerife where 200 young chimpanzees are kept as a profitable sideline by photographers who know that holidaymakers like to have their pictures taken with one of these animals.

Primates as zoo animals

Prior to the 1st century A.D. primates were kept with relative success in the zoological gardens of Egypt and China as well as in Montezuma's realm in Central America in the 15th century. These animals were intended to enhance the prestige of the rulers and privileged classes, just as they were later on in Europe. Thus it was not until after the French Revolution of 1789 that in Paris the Jardin des Plantes, the first zoological garden, and in the following year other gardens too, were thrown open to the public so that people could admire, amongst other animals, the strange man-like creatures from the faraway tropics that were still shrouded in mystery. It was in Paris too that the French naturalists G.L.L.Buffon (1707–1788), J.B.Lamarck (1744–1829) and E.Geoffroy de Saint-Hilaire (1772–1844) put forward their early controversial evolutionary theories in which the aye-aye played a part. The era of modern zoological gardens did not begin until the mid-19th century. Today all zoos run along up-to-date lines are primarily scientifico-biological educational and research establishments, whereby the primates attract the greatest interest. The collections of primates provide a survey of their manifold aspects enabling comparisons to be made with humans, as well as ecological and ethological studies and observations.

The keeping and care of primates in zoological gardens is based on present-day biological knowledge. For instance, apes are kept in large group rooms with a surface area of between 100 and 140 square meters behind safety glass in purpose-built housing with facilities for partition and separation as well as diversified interior equipment. There are also double or alternative rooms with a unitized system of interchangeable equipment so that a temporary change of scenery can be effected. TV programs with plenty of action are a more or less successful way of checking stereotype movement and abnormal self-absorption. A diet based on sound biological principles and specific indoor climatic conditions contribute toward health and longevity. Spacious and airy primate houses equally conform with modern zoological and biological stipulations. Only forty or fifty years ago tuberculosis, parasites and gastro-enteric complaints were the chief causes of death among zoo primates. Large gibbon islands allow these very accomplished acrobats plenty of space to swing, grab branches and swoop off again to the delight of the spectators. Spider monkeys too can be given a similar suitable environment. By keeping colonies of baboons on rocky ground it is possible to gain an impressive picture of these animals' pronounced social behavior patterns. This kind of large open-air enclosure is also very suitable for macaques, while magots are in some places kept in a "monkey forest," an extensive woodland and meadow area enclosed by a live-wire fence with a circular path inside, so that people can see the monkeys without being separated by meshing. Even the squirrel-sized marmosets and tamarins require large cages for long survival and for breeding purposes. Using modern methods successful efforts have been made to keep apes in Basle, Frankfurt on Main, Stuttgart, Atlanta, Cincinnati, Chicago (Lincoln Park), Colorado, Memphis, Philadelphia, Dresden, Berlin. H.Hediger (1977) summarized the desirable conditions for the keeping of primates and other animals in zoos in the following words: "Accommodation of the animals in an appropriate biotop in territories and social groups conforming as far as possible to natural conditions and a diet approximating to their normal nutrition . . . Furthermore a limitation of the number of species through the reduction of individual specimens, with preference given to breeding groups of the endangered species, and all these factors in a landscape designed for human recreation. Making due provision for the proper care of the animals, there should be optimal facilities for popular education and scientific research."

The need for protective measures

Generally speaking, primates are not coveted trophies. Nonetheless, earlier on big-game hunters killed the great apes, partly out of ignorance. The primates with attractive colored fur, for instance the guerezas, are still hunted today although this practice has been largely restricted. In some regions monkey flesh is often eaten. When large numbers of these animals damage plantations they are shot. Nature pays a high tribute every year in terms of macaques, guenons, capuchins, marmosets and tamarins which are needed for laboratory purposes especially by the pharmaceutical industry. In comparison with the overall figure, the supply by traders of primates for zoos and as domestic pets only represents a small proportion. Despite all the other urgently needed restrictions, more attention in the form of effective controls should be paid to the losses incurred in the process of capture and transport. Nowadays primates are nearly always transported as air freight so that they reach their destinations from the most remote corners of the earth in two or three days at the most. But air cargo costs are relatively high and for reasons of economy the animals are sent in very small containers that frequently harm them. Hence at large airports with a great deal of freight transshipment there are animal care centers where during stopovers they can be attended to and given water—their principal need.

During recent years a whole number of countries have established new animal reserves in which primates too can be found. These include Sierra Leone (chimpanzees), Gambia and Ghana. The great significance of research centers in these protected areas is generally admitted. The same applies to the rescue centers for captured orang utans on Sumatera and for chimpanzees in Africa. So far it has been proved beyond a doubt that depleted primate herds increase in nature conservation areas provided the original environment is not destroyed. As a result of the extensive destruction through herbicides of cultivated and forest areas in Vietnam the primate population was decimated too. The same happened in Uganda where especially in 1979/80 during the military conflicts marauding soldiers did much damage to the animal conservation areas and to Ruwenzori, Kabalega and Kidepo, the three large national parks. In 1980 the Tanzanian government decided to enlarge the Arusha National Park. The Mahali National Park to the southeast of Lake Tanganyika is to be revived. Research workers who have lived there for many years estimate that 2,000 chimpanzees inhabit the Mahali Mountains. South of Lake Victoria the Burugi National Park covering an area of 3,400 square kilometers is also to be reconstructed. It is the range of guenons and colobus monkeys.

In Japan deformed primates have alarmed environmentalists in recent years. In 22 of the 61 primate herds investigated, many animals with shortened and crippled or even missing limbs as well as badly damaged internal organs were found. A half-tame herd of macaques on Awaji that was chiefly fed by humans displayed extremely severe effects. The deformations were caused by agrarian chemicals, for instance preservative agents for apples and citrus fruits, contained in the primates' food supply. An alarming threat to primates everywhere is the increasing encroachment upon their natural habitats. Some years ago even part of the Serengeti was turned over to cattle farming on a chiefly extensive basis. Similarly, the small protected area for lion-tailed monkeys (only 22 square kilometers) in Southwest India—the total number of these animals is estimated to amount to only 825—has still not undergone the urgently necessary extension.

There is no lack of demand for guenons from Kenya, but neglected nature conservation has led to a much reduced export of these animals.

The number of Madagascar's prosimians also continues to be jeopardized. According to F. Steiner the island with an area of 587,000 square kilometers only retains 50,000 square kilometers of its original vegetation. There remains a bare 10 percent of the primary vegetation, whereas more than 80 percent consists of barren grasslands as secondary vegetation which hardly or only partially corresponds to the prosimians' natural biotop. Madagascar is a typical example of the effects of unrestricted inroads on the natural tropical forest. In recent years many countries have experienced a growing demand for tropical timber. The felling of timber on an industrial scale has reached alarming proportions in the rain forests of Africa, South and Southeast Asia and South America. These are the regions where the great majority of primate species live. The original Food and Agricultural Organization (FAO) estimate of an annual loss of about 6 million hectares of rain forest was shown to be very wide of the mark when in 1981 an American scientific institution analyzed satellite photos: the approximate annual loss is 20 million hectares! If these appalling inroads continue unchecked the last giant tree of the tropical forest will be felled in about 45 years' time. According to estimates made by the UN Economic Commission for Latin America the crisis will assume the dimensions of an ecological tragedy in 1995.

In the previously mentioned large Gunung Leuser Reservation in Indonesia on northern Sumatera about 2,000 orang utans have been given an undisturbed forest habitat. A protected area covering 550 square kilometers on Siberut, one of the Mentawai Islands, is to save the four species found only in this region, the dwarf siamang, the Mentawi leaf monkey, the Pageh pig-tailed macaque and the Pageh pig-tailed langur from extinction. Having realized the danger to its fauna, the government in Kuala Lumpur decided to establish national parks for each state in Malaysia and this will promote the survival of many primate species. Efforts to protect certain species have already existed for many years. In a number of countries, principally in Europe, there are well-established nature conservation laws on a national or regional level which exclusively concern the native fauna, but rarely or only seldom apply to primates. The first continental agreement which also included primates (gorillas, all Madagascar primates) was the International London Convention for the Protection of African Flora and Fauna which came into force in 1936. It is today substantially observed, albeit with amendments, by most states of this continent. After 1945 the International Union for Conservation of Nature and of Natural Resources (IUCN) came into being with its headquarters in Glaudes (Switzerland). In order to introduce effective protective measures the Survival Service Commission collected all available information about the numbers of existing species. The list of mammals published in 1966 contains 277 rare and endangered species or subspecies. About 70 of these were listed on red paper which subsequently became the *Red Data Book*; it chiefly provides information about species and subspecies threatened with extinction and includes many primate species. During the past 25 years in particular, world-wide animal protection has broadened its foundations. The World Wildlife Fund (WWF) with its headquarters in Glaudes, the International Society for the Protection of Animals (ISPA), London, the Fédération Mondiale pour la Protection des Animaux (FMPA), Anvers, the American Committee for International Wild Life Protection, New York, UNESCO, Paris, FAO, Rome, the International Federation for the Protection of Primates, Berkeley (USA) and others are engaged in this important mission. Amongst the organizations operating at a regional level, special mention should be made of the Zoologische Gesellschaft von 1858, Frankfurt on Main with its campaign to help the threatened fauna, and financial assistance similar to that rendered by the WWF for the return of smuggled chimpanzees to Africa.

It is our duty to make the zoos along with the wild life protected areas refuges and successful breeding centers for all endangered primates too. Good results that give rise to optimism have been achieved with macaques, lion-tailed and patas monkeys as well as guenons, mandrills, baboons, guerezas, capuchins, squirrel monkeys, ring-tailed lemurs, mongoose lemurs,

Approximate present geographical distribution of prosimians, monkeys, and apes.

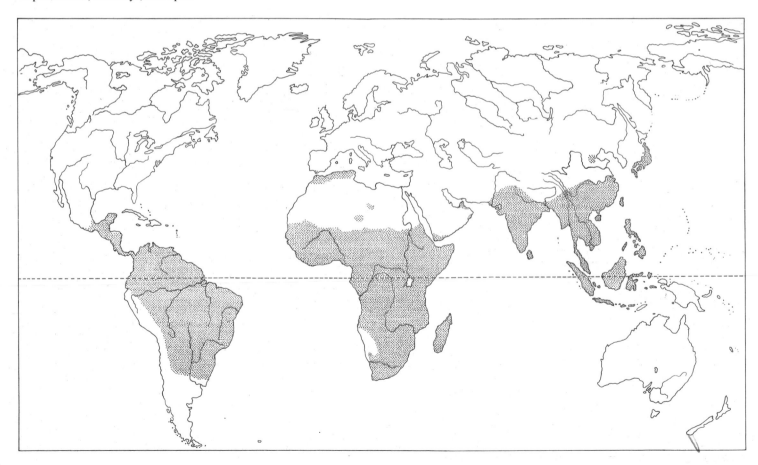

etc. and equally in the case of white-handed gibbons, chimpanzees, orang utans and here and there with gorillas. For instance, following the example of Basle, where gorillas of the second zoo generation have been bred, Frankfurt on Main reported the tenth gorilla birth—all hand-reared—in 1980. In addition the bonobo "Margit" gave birth to her fifth child that year. The Cologne Zoo is internationally famed for its remarkable successes in the rearing and breeding of prosimians, New World and leaf monkeys. There are other positive examples of this kind. Through breeding their own stock the zoos can meet their primate requirements in the first place and later on hand over animals to the wild life reserves in their countries of origin. The International Union of Directors of Zoological Gardens (IUDZG) staunchly supports these protection efforts. For example, no wild animals captured in their original homelands are purchased. The incentive and high profits of the illegal trade are thus heavily curtailed. The international studbooks for the species threatened by extinction are a valuable achievement. The annual entries give an exact survey of breeding processes and the respective number known to exist. Studbooks are kept at the Washington National Zoological Park for the lion marmoset; at the San Diego State University for the douc langur; in Woodland Park Zoo for the lion-tailed monkey; in the Zoological Society of San Diego for the orang utan; in the Frankfurt on Main Zoological Gardens for the gorilla; and in Antwerp Zoo for the bonobo.

In the highly informative annual *International Zoo Yearbook* (London) the number of rare animals in captivity and their offspring reared by human hand is listed. A very effective and decisive step toward the protection of endangered fauna and flora was the Washington Convention of March 1973 on International Trade in Endangered Species of Wild Fauna and Flora which has so far (1981) been signed by 62 countries.

The threatened species are listed in three differentiated appendices:

Appendix I includes all species threatened with extinction. The capture and trade in these species is subject to particularly strict regulation and can only be authorized in exceptional circumstances.
Appendix II includes all species which may be threatened with extinction unless trade with specimens of these species is subject to strict regulation, and other species which must be subject to regulation in order that trade in them may be brought under effective control.
Appendix III includes all species which any signatory identifies as being subject to regulation within its jurisdiction for the purpose of preventing or restricting exploitation, and as needing the cooperation of other parties in the control of trade.

All prosimians and primate species are included in either Appendix I or II according to the degree to which they are endan-

gered. Should changes either of a positive or negative character occur in the number of a species in the course of time, the species concerned can be transferred from one appendix to the other. Appendix I includes the following species and groups (as of June 6, 1981):

Lemuridae: all species.
Indriidae: all species.
Daubentoniidae: the only surviving species.
Callimiconidae: the only surviving species.
Callithricidae: white-eared marmoset, buff-headed marmoset, pygmy marmoset, lion marmoset, golden-headed tamarin, golden-rumped tamarin, pied tamarin, white-footed tamarin, cotton-headed tamarin, Geoffroy's tamarin.
Cebidae: red uakari, scarlet uakari, black-headed uakari, black uakari, white-nosed saki, red-backed squirrel monkey, mantled howler monkey, woolly spider monkey, black-handed spider monkey, Panama spider monkey.
Cercopithecidae: Tana agile mangabey, lion-tailed monkey, Diana monkey, drill, mandrill.
Colobidae: entellus monkey, capped langur, Gee's langur, Mentawi leaf monkey, douc langur, Pageh pig-tailed langur, proboscis monkey, red colobus, Zanzibar colobus.
Hylobatidae: all species.
Pongidae: all species.

Framing and acceding to this Convention is the first step: its universal enforcement and effective controls are, however, the decisive phase. There are unfortunately still factors that encourage international animal smuggling and trade with fur, hides and other parts of the bodies of protected animals. Even today in Sierra Leone, for example, large numbers of chimpanzees are captured and taken out of the country without registration. There are still plenty of customers for them. More and more frequently after confiscation such illegally imported animals are brought back to rescue centers in Africa. In most of the regions where wild animals are caught, although the protection of the relevant species is a matter of national prestige, a true appreciation of what is at stake is lacking. This complicates the work of the customs and management authorities in Europe. Even today, gorillas, for instance, are sold for 70,000 Marks per animal, bonobos for 12,000 Marks and chimpanzees for 5,000 Marks by unscrupulous traders.

The International Society for Primatology, which also makes great efforts to protect primates, met in Bengalur (India) for its 7th Congress in 1979. According to the last census, half of the approximately 180 species of primates is threatened with extinction. In order to reverse this process even greater efforts must be made to study their living conditions. Precisely because it is a necessity for primates to be available, for medical purposes too, their protection is a universal concern. This protection—according to M. Kavanagh and D. Chivers (1979)—poses an almost insoluble problem for the environmentalists so long as it is not possible to put large stretches of tropical forest under absolute nature protection. In Bangladesh, for instance, there are now only three large compact forest areas, none of which is protected. The large numbers of hanuman monkeys and other langurs that used to live there are now probably extinct, more especially because they are not regarded as sacred animals as in India. So far the evergreen tropical rain forests of Malaysia have provided better conditions in the form of a relatively favourable habitat for many species of gibbons, macaques, langurs and prosimians. Despite the threat of deforestation—only about half the area has tree cover—it is hoped to maintain 45 percent forested territory since the catastrophic effects of an unrestricted exploitation of irreplaceable natural resources have been realized. As already mentioned, there is an organized system of wild-life protection and national parks as well as the necessary research staff and keepers here. The effective protection of species should not be just left to individual organizations, international conventions and regional legislation; the population in general should identify itself with this cause. Human beings alone are in a position to assume responsibility for nature and life itself and to cherish and protect it.

Description of the Families, Genera and Species

The Order of Primates

Prosimians
(suborder Prosimiae)

The primates made their appearance as a distinct group of mammals at the start of the Paleocene period, about 60 million years ago. *Plesiadapis*, the earliest known primate, has been found in Europe and North America.

The Paleocene primates were small, short-legged animals with long tails and snouts, indicating a well-developed sense of smell. Their vision was limited since their little eyes were still placed on the sides of the head. Each finger and toe ended with a claw. These shrew-like creatures lived in the trees. Adaptation to an arboreal life is in fact the common feature shared by all living and extinct non-human primates.

Through the study of fossil remains and of existing species we know that in the course of many millions of years the prosimians evolved from the insectivores; modern mammal classification therefore places the primates next to the insect-eaters on the ladder of evolution.

The word "prosimian" or "pre-monkey" indicates its intermediate position in the transition from insectivore to primate. The many prosimian varieties are classified into four infraorders with seven families: tree shrews, lemurs, indris, aye-ayes, lorises, bush-babies, and tarsiers. Some species closely resemble the insectivores, others are more simian. In contrast to the insectivores that are largely dependent on their sense of smell, the prosimians are guided mainly by their eyes. Their snouts are less pronounced than those of the insectivores. An almost universal characteristic of the prosimian is the bare moist glandular skin surrounding the nostrils (rhinarium). Many mammals possess this feature, but the tarsiers and all monkeys and apes have lost it. The sight control centers in the brain are better developed than those controlling the sense of smell. The eyes have moved more to the front so that the fields of vision overlap to produce stereoscopic sight. For a primate living in the trees it is essential to be able to judge distances when jumping or climbing. With the exception of the tree shrew and with certain variations the dental formula is $\frac{2 \cdot 1 \cdot 3 \cdot 3}{2 \cdot 1 \cdot 3 \cdot 3}$. The prosimians have thick coats and usually long tails which are not prehensile. The long, slender fingers and toes of the prosimians are furnished with nails except for the second toe. Whereas other mammals that live in the trees hold on to the branches by digging in their claws, the prosimians grasp them with fingers and toes. The soft broad pads on their digits resemble those of monkeys. Flexible thumbs and big toes make grasping possible and lessen the risk of falling.

The prosimians' eyes are very large, especially in the case of the nocturnal species. Their sense of hearing is highly developed. They often have strikingly large outer ears. The types that specialize in jumping have longer tarsal and navicular bones. The still primitive form of a bicornuate uterus is typical for the prosimians. There is generally a fixed reproductive cycle. One to three offspring are born at a time and are either deposited in a nest or travel around with their mother, clinging to her abdomen.

As tree-dwellers, the prosimians spread over North and South America, Europe, Africa and Asia. So far the fossils of more than 60 extinct prosimian species have been found in various parts of the world. Most of these remains belong to the Lemuriformes and Tarsiiformes groups.

Gradually the monkeys displaced the prosimians. In Africa and Asia certain prosimian species are still found that, being nocturnal have not to cope with a confrontation with the monkeys which are only active during the day. One prosimian group found a refuge on Madagascar where several families could evolve without being threatened by monkeys or beasts of prey.

Infraorder Tupaiiformes

Tree Shrews (family Tupaiidae)

At a cursory glance it would seem that the tree shrew with its primitive and uncomplicated anatomy belongs to the squirrel-like insect eaters. And formerly this animal, which inhabits the tropical parts of Southeast Asia and its adjoining islands, was in fact classified as an insectivore.

After studies made by Le Gros Clark and H. Sprankel it was placed at the root of the primates' family tree, hence forming a link between the insectivores and the monkeys and apes. This was because its eye-sockets are encircled by bone and its tongue is underlain by an elastic, partly cartilaginous sub-tongue; and also because of certain features in the structure of its reproductive organs, muscles and skull. Not all authorities consider that this classification is justified, and some taxonomists, bearing in mind this primitive animal's reproductive system and embryonal development, think that it should again be separated from the primates.

Tupaia is derived from the Malay word "tupai" which is used as a designation for all squirrel-like animals in their tropical habitat. These mobile little tree-dwellers, measuring between 10 and 23 cm, have slender bodies and long snouts. Their ears are naked and they have long bushy tails. Their fur is thick, short and soft. Each of the fingers and toes on their little hands and long feet ends with a claw. Their thumbs cannot be opposed to the fingers so they have not got a true grasp.

Apart from insects the tupaias eat fruit, leaves, birds, eggs, and even small rodents.

The family is classified into two subfamilies with six genera, 47 species and about 100 subspecies.

Various phases in the movements of a tree shrew (*Tupaia* spec.) (after Jenkins).

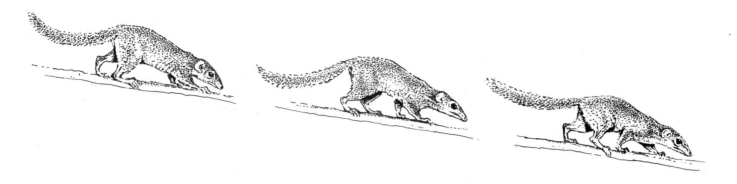

Bushy-tailed Tree Shrews (subfamily Tupaiinae)

Tree Shrews (genus *Tupaia*)

Common Tree Shrew *(Tupaia glis)*. Of the 14 tupaia species the common tree shrew is the best-known and most extensively studied. The upper side of the body is dark with irregular gray, brown or black patches, the underside white or buff. Head and body length ranges from 14 to 23 cm; the bushy tail is as long as the body. The hind legs are somewhat longer than the forelimbs and the weight is about 170 grams. The common tree shrew is found in Farther India and South China.

It lives alone or in pairs and is active during the day. It chiefly eats insects and other invertebrates, but also eggs, fruit and leaves. After gestation lasting from 41 to 50 days between one and four (an average of two) young are born. Their birthplace is a hole in a tree padded with foliage. They are naked and blind at birth and weigh from 11 to 15 grams with a head and body length of 6 cm. According to Sprankel, the young are suckled two or three times a day and according to Martin only once in 48 hours. They open their eyes on the 15th day and on the 20th day after birth start to take solids. Their growth is completed after three months and they are sexually mature at four months.

These prosimians rely very much on their sense of smell to guide them. The males mark their territory with scent from their chest glands which exude an oily, pungent and sticky secretion. They also mark their territory with frequent drops of urine. Sometimes these animals, like the bush-babies, urinate on the inner surface of their hands and feet. They use a wide range of sounds to communicate with each other. In Europe these tree shrews were successfully bred for the first time at the Cologne Zoo and at the Frankfurt Max Planck Institute for Brain Research. The zoological gardens in Dresden and Leipzig have scored outstanding successes in raising these tree shrews artificially. Further tupaia species that live in Southeast Asia are:

Günther's tree shrew *(Tupaia minor)* on Sumatera; the **mountain tree shrew** *(Tupaia montana)* on Kalimantan and the **Javanese tree shrew** *(Tupaia javanica)*.

Greater Tree Shrews (genus *Tana*)

Tana or Greater Tree Shrew *(Tana tana)*. A forest-dweller, this tree shrew lives more on the ground than in the trees. Its main sources of sustenance are worms, insects and some kinds of fruit. Measuring between 16 and 24 cm, it has a tail as long as its body and weighs from 160 to 260 grams. Its fur is dark red and brown with a black stripe over its back. The female has two pairs of mammary glands. Gestation lasts 44 days and the young weigh 9 to 11 grams at birth. This species lives on Kalimantan, Sumatera and the adjoining islands.

Indian Tree Shrews (genus *Anathana*)

Elliot's Tree Shrew *(Anathana ellioti)*. The only species, it is active during the day and lives in the tropical rain forests of southern India.

Philippine Tree Shrews (genus *Urogale*)

Everett's Tree Shrew *(Urogale everetti)*. The sole species of this genus, it lives in the dense forests of the Philippines. Active during the day, this tree shrew is unusually large, has a long snout and a short-haired tail. It lives on insects, small lizards, birds' eggs, and frogs.

Smooth-tailed Tree Shrews (genus *Dendrogale*)

These small tree shrews have very sleek fur on their tails. Their faces display fairly conspicuous markings but they have no shoulder stripe. Active during the day, the *Dendrogale* occur in various areas of South China and on Kalimantan. There are two species: *Dendrogale murina* and *Dendrogale melanura*.

Pen-tailed Tree Shrews (subfamily Ptilocercinae)

Pen-tailed Tree Shrews (genus *Ptilocercus*)

Pen-tailed or **Feather-tailed Tree Shrew** (*Ptilocercus lowii*). The only known species lives in the trees of the Southeast Asian virgin forests. It has a dark stripe running sideways from the snout. Its large eyes and big pliable ears indicate nocturnal or twilight activity. A characteristic feature is the naked scaly tail, tufted at the end which, as an agile climber, it uses as a counterbalance. On the ground it moves by jumping, keeping its tail in a vertical position. When asleep, it curls its tail round its body in such a way that the tufted end hangs over its head like a fan. In the wild the pen-tailed tree shrews normally live in pairs, although in the daytime four animals have been found in a nest. They mainly feed on insects. Little is known about their habits and they have never lived in captivity outside their homeland. The two known subspecies are:

Ptilocercus lowii lowii on Kalimantan and the mainland variety *Ptilocercus lowii continentis* in Malaya.

Infraorder Lemuriformes

The range of the lemurs is restricted to Madagascar and the Comoro Islands. Like Australia, Madagascar is regarded as a "zoological treasure-chamber," and French naturalists in the 19th century called it "le paradis des naturalistes." About forty million years ago, in the Oligocene period, the Mozambique Channel came into being, separating Madagascar from the continent of Africa by about 390 kilometers. Hence all the mammals that later migrated into Africa, including the monkeys, are missing here. On the other hand groups of mammals that have long been ousted by other animals on the African mainland have survived and evolved on Madagascar.

There still exist about 20 lemur species belonging to 11 genera. The fossil remains of prosimians include not only small species but also some that were comparatively large. One of these was the huge *Megaladapis*, the size of a donkey. Some of these varieties, including the larger forms, coexisted with still living species and only became extinct during the past 3,000 years. Fourteen large prosimian species disappeared in this way. Today the majority of Madagascar's indigenous mammals are severely threatened by changes in the vegetation cover. Large parts of Madagascar, particularly the west and southwest, have become almost barren. Extensive tracts of the original tree cover have already disappeared. If the remaining forests are not protected against everything that threatens them, Madagascar will also lose its indigenous primates. So far 12 nature reservation zones have been established in various parts of the country. Unfortunately it is hardly possible to ensure the observation of the protective regulations. Madagascar is forced to rely on international support for the future survival of its unique natural features.

Theoretically speaking, all Madagascar primates enjoy full protection under the terms of the London Convention of 1936 (see page 23). A strict export ban has existed for many years so that most of the species are seldom seen in zoos. The survival of the genera *Phaner*, *Hapalemur*, *Indri*, and *Daubentonia* is today particularly endangered.

True Lemurs (family Lemuridae)

French explorers gave the extremely active and engaging Madagascar lemurs their name, although they are not in the least weird or ghostly like the Roman *lemures*, the spirits of the dead. Probably the explorers were influenced by the lemurs' moaning and often unnerving calls or their nocturnally luminous eyes.

The lemurs are between mouse and cat size, have round heads with more or less pointed snouts, large eyes and small ears, thick fur and bushy tails. They have flat nails; on the second toe they still have a claw which, together with the incisors, is used for scraping purposes. The Madagascar lemurs that live in troops can often be seen grooming each other. Their strong hind legs help them to leap long distances. There are varieties active either by day or by night, some that live singly, others in troops. There are species which are omnivorous, others that live on fruit and some which chiefly live on insects. Scent and smell play a more important part in their lives than sight. Their reproductive organs are still primitive so there is no menstruation. Birth occurs at a certain season of the year.

Dwarf Lemurs (subfamily Cheirogaleinae)

Mouse Lemurs (genus *Microcebus*)

Nocturnally active, these tree-dwellers are the smallest members of the lemur family. In appearance they resemble bush-babies, but their large ears cannot be folded together. They are insectivores but also eat fruit and sticky tree sap.

Lesser Mouse Lemur (*Microcebus murinus*). This brown or gray prosimian measures between 11 and 13 cm, its tail is 12 cm long and it weighs from 50 to 60 grams. A characteristic feature is the light stripe on its short pointed snout. During the rainy season when food is abundant, fat is stored in cushions around the hind legs and base of the tail. In the dry season it goes into a state of suspended animation, living on its fat deposits. Mouse lemurs use twigs and leaves to build their nests in which the female gives birth to between one and three young after gestation lasting from 59 to 62 days. The offspring are from 3.7 to 5 cm long at birth, the tail measures 3 cm and their weight ranges from 2.7 to 4.3 grams. The body is covered with grayish brown fur and their eyes open after four days. The mother transports her young by gripping them with her teeth. After 12 days the young animals can jump and at 15 days start to climb. They are suckled for 45 days and become independent at two months.

Agile climbers and jumpers, these tree-dwellers are found in the wet upland forests, in reedy zones and in the euphorbia bushes of the dry grasslands along the east and west coasts of Madagascar. Zoos with modern houses for nocturnal animals exhibit growing numbers of these dwarf primates. There are instances where they have lived in captivity from six to ten years. There are two subspecies:

Microcebus murinus murinus, West and Southeast Madagascar, and *Microcebus murinus smithii*, Northeast and East Madagascar.

Coquerel's Mouse Lemur (*Microcebus coquereli*). This grayish brown animal about the size of a European squirrel is found in southwestern and northwestern Madagascar. It has a round head, a short pointed snout, lyrge eyes, fairly small ears and a bushy tail about 28 cm long. It is still uncommon in zoological gardens and is today regarded as an endangered species.

True Dwarf Lemurs (genus *Cheirogaleus*)

The three species of dwarf lemurs have thick woolly coats in which their partially furry ears are hidden. Their body temperature drops when the air grows colder. Their metabolism stows down and they sink into a state of suspended animation, subsisting on the fat deposits they have accumulated on their hind legs and at the base of their tails. They feed mainly on fruit, but also eat insects. The entire *Cheirogaleus* genus is today regarded as acutely threatened.

Greater Dwarf Lemur (*Cheirogaleus major*). These brownish red or gray animals live in the wet forests of eastern Madagascar. With a body length of from 19 to 27 cm, a tail measuring from 16.5 to 25 cm, they weigh 350 grams. They build nests of leaves in the tree-tops and several animals spend the day there together. In the cold season this prosimian has spells of lethargy when its body temperature sinks almost to the level of its environment. In some regions it is said to hibernate for weeks at a time in tree trunks. This lemur marks its territory with urine or excrement which it rubs on the surface of the branches.

Gestation takes 70 days after which two or three young weighing from 15 to 20 grams are born whose eyes open on the second day after birth. This lemur feeds on fruit, blossoms, insects, and small vertebrates. The two known subspecies are:

Cheirogaleus major major, Southeast Madagascar, and *Cheirogaleus major crossleyi*, West Madagascar.

Fat-tailed Dwarf Lemur (*Cheirogaleus medius*). This rat-sized species is found in the dry forests of western and southern Madagascar. Weighing 250 grams, its fur is brownish red or gray and its underside is almost white. Its chief features are the bushy tail, 20 cm in length, and large close-set eyes. In the dry season or when the weather is cold it hibernates for several weeks, living on its fat deposits.

Hairy-eared Dwarf Lemur (*Cheirogaleus* [*Allocebus*] *trichotis*). This lemur has a total length of only 30 cm, more than half of which is taken up by the tail. Characteristic features are thick tufts of fur on the ears and narrow, ridged and pointed nails. Detailed information about this lemur that lives in the upland forests of eastern Madagascar is not available since so far only four specimens have been found.

Phaner (genus *Phaner*)

This genus differs from *Cheirogaleus* by virtue of its elongated and slanting upper incisors. On each side of the canine teeth in the upper jaw there is a premolar resembling a canine tooth. Large, naked and rounded ears are also characteristic.

Fork-marked Dwarf Lemur *(Phaner furcifer)*. In size and appearance similar to the greater dwarf lemur although its legs are longer and its ears larger, the only known species lives in the north and west of Madagascar. A forest-dweller, the "waluwy" as it is called by the local people, was named after the dark line along the center of its back which forks at the top of the head, the branches merging into the black rings round the eyes. This mobile, nocturnal lemur lives alone or in pairs in low-lying forest areas, is very agile and can jump a distance of several meters. With its long sharp claws it can grip the bark of large tree trunks. The males have a naked patch of glands below the neck and use the scent to mark their territory. This lemur's diet includes among others nectar, gum resin, and insect secretions.

During the rainy season one young is born that is first of all deposited in a hole in a tree and then carried ventrally by the mother. Later on it rides on her back. Further details are not known. This species is rarely found in captivity and has never been bred in zoos.

Typical Lemurs (subfamily Lemurinae)

True Lemurs (genus *Lemur*)

These lemurs have fox-like faces and a lively disposition. They live in the trees and are active during the day and at dusk. Of medium size, they weigh about two kilos and are socially minded, living in groups. The hind limbs are longer than the fore limbs, the tail is longer than the body. The six species of this genus live on Madagascar and the Comoro Islands in both dry and wet forests and also occur at high altitudes. They mainly eat fruit, leaves, and blossoms. They are the least specialized of Madagascar's prosimians.

Brown Lemur *(Lemur fulvus)*. Individual specimens vary a good deal in their fur coloring. Their classification is a matter of controversy. Entirely arboreal, they are found in all Madagascar's forested zones with the exception of the extreme south. There are six subspecies:

Lemur fulvus fulvus. Individuals vary widely in their fur coloring. They live amicably in troops of about twelve animals on the east coast of Madagascar.

Lemur fulvus rufus. The male's forehead is red. The female has two white patches over the eyes. In the dry season they live mainly on leaves, pods, stems, blossoms, and the sap and bark of *Tamarindus indica.* They live in troops numbering from 4 to 17 animals in the dry zones of western Madagascar.

Lemur fulvus albifrons. The white fur round the face of the male stands out in contrast to the black mouth. The female has drab grayish brown fur. Gestation lasts from 130 to 133 days; one or two offspring are then born which are carried ventrally by the mother. Found in the forests of northern Madagascar.

Lemur fulvus sanfordi. The male has short white ear tufts. Found in the Amber Mountains in the north of Madagascar.

Lemur fulvus collaris. Fur grayish brown, underside lighter. The female's head is anthracite colored while the male's is black. Found in southwestern Madagascar.

Lemur fulvus mayottensis. The fur coloring varies a good deal. Characteristic features are the side-whiskers, the male having a more prolific growth. Gestation lasts about 130 days after which one or two young are born and then carried ventrally by the mother. Only found on Mayotte Island (366 square kilometers) to the northwest of Madagascar.

Black Lemur *(Lemur macaco)*. The following subspecies are known:

Black Lemur *(Lemur macaco macaco)*. The male has deep black fur with ear tufts the same color. The female's fur is yellow to rusty brown, cheeks and ear tufts are white. Called "akumbo" by the local people, these lemurs mark their territory very frequently, using both the anal glands and those over their wrists. They live socially in groups numbering from 6 to 15 animals. The number of males always exceeds that of females although the latter dominate the group. Gestation lasts from 127 to 130 days and one grayish black baby is born each time. When two weeks old the young animals, which are carried around ventrally by the mother from the time of birth, begin to climb on to her back and become independent at six months. The animals communicate by means of deep grunts; when alarmed they utter shrill cries. In the evening they strike up an unearthly wailing concert. They live in the tallest trees of the dense, impenetrable forests of northwestern Madagascar.

Lemur macaco flavifrons. The males are brownish in color and their fur forms a short crest at the top of the head. The female's fur is a pale reddish brown on the dorsal side. Northwestern Madagascar.

---

5 Common Tree Shrew *(Tupaia glis)*. Although the tree shrews still have much in common with the insectivores, they are nowadays placed at the root of the primate family tree. (Prague Zoological Gardens)

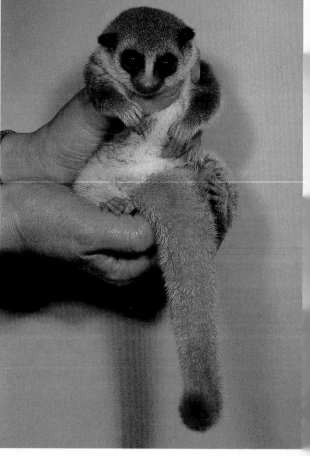

6/7 Lesser Mouse Lemurs
(Microcebus murinus).
The mouse lemur is one of the most
common lemur species on Madagascar.
(Cologne Zoological Gardens)

8 Fat-tailed Dwarf Lemur
(Cheirogaleus medius).
The layer of fat measuring up to 2 cm
deposited in the tail enables the animal to
survive the cool and dry season.
(Cologne Zoological Gardens)

9 *Lemur fulvus collaris,*
a subspecies of the typical lemur, with young.
(Cologne Zoological Gardens)

10 Black Lemur *(Lemur macaco).*
Large ear tufts and a beard-like fringe
of hair on the cheeks are the black lemur's
distinguishing features. (Cologne Zoological
Gardens)

11 *Lemur fulvus melanocephalus.*
Female with young. (Cologne Zoological Gardens)

12 *Lemur fulvus albifrons.*
Only the male has this conspicuous ruff of
white fur round its face. (Cologne Zoological
Gardens)

13/14 Crowned Lemurs *(Lemur coronatus).*
This lemur was formerly regarded as a
subspecies of the mongoose lemur. Because
of its many divergencies it was classified
as a distinct species in 1975.
Left: Young animals. (Cologne Zoological Gardens)

15 *Lemur fulvus albifrons.*
The gleaming white fur on the crown, cheeks
and beard gives the male of this subspecies
an attractive appearance. (Cologne Zoological
Gardens)

16/17 *Lemur fulvus mayottensis.*
Below left: A two-months-old hand-reared
specimen. (Cologne Zoological Gardens)
Right: The male has a more prolific growth
of whiskers. (Cologne Zoological Gardens)

18 Black Lemur *(Lemur macaco)*.
The coloring varies according to sex.
(Cologne Zoological Gardens)

19 Mongoose Lemur *(Lemur mongoz)*.
Male. (Cologne Zoological Gardens)

20 Crowned Lemur *(Lemur coronatus)*.
Like other prosimians the mother carries her
baby ventrally. (Cologne Zoological Gardens)

21–24 Ring-tailed Lemurs (*Lemur catta*).
Above: Ring-tailed lemurs are sociable
animals and form troops of up to 22 members—
the largest amongst the lemurs. When moving
on all fours the ring-tailed lemur holds its
long tail erect with the tip drooping backward
as a signaling device that aids contact
within the group. (Cologne Zoological Gardens)
Above right: Liberec Zoological Gardens
Below right: The ring-tailed lemur's typical
and favorite posture when taking a sunbath.
The erect body with outstretched arms forms
a fascinating picture. (Prague Zoological
Gardens)

Opposite page:
The long fang-like canine teeth and the
glandular surface above the right wrist are
clearly visible. (Cologne Zoological Gardens)

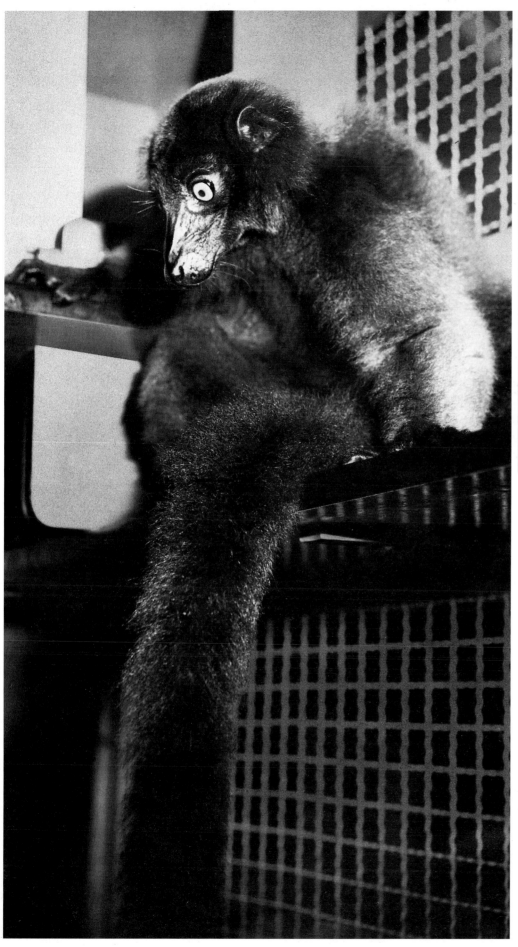

25 Ruffed Lemur *(Varecia variegata)*.
The ruffed lemur's black and white coat
provides excellent camouflage in the rain
forest's diffused play of light and shade.
(Cologne Zoological Gardens)

26 Red Ruffed Lemur *(Varecia variegata ruber)*
(Cologne Zoological Gardens)

27 Gray Gentle Lemur *(Hapalemur griseus)*.
When threatening, this lemur stands erect,
laying back its ears and drawing its tail
several times between the inner surface of
the right or left lower arm.
(Cologne Zoological Gardens)

Overleaf:

28 Mongoose Lemur *(Lemur mongoz)*.
The markings of these dark brown prosimians
vary according to sex. The male animals have
red cheeks and a reddish nape, while the
females have white cheeks and throats.
(Prague Zoological Gardens)

29 Gray Gentle Lemur *(Hapalemur griseus)*.
Gentle lemurs mainly live on bamboo shoots
and leaves. Their teeth, including the pointed
incisors, are adapted to this specialized diet.
(Cologne Zoological Gardens)

30 Demidov's Bush-baby *(Galago demidovii)*.
It owes its name to the cry it utters which
sounds like a loudly wailing infant. (East Africa)

31/32 Slow Loris *(Nycticebus coucang)*.
The second toe of both the potto and the slow
loris is furnished with a long and strong
claw used for scraping. (Cologne Zoological
Gardens)
Opposite page:
With its round head, large dark close-set
eyes and thick woolly fur the slow loris looks
like a teddy bear. (Dresden Zoological Gardens)

33 Ruffed Lemur *(Varecia variegata)*
(Cologne Zoological Gardens)

34 Red Ruffed Lemur *(Varecia variegata ruber).*
A beige spot on its nose is typical for
this ruffed lemur subspecies. (Cologne
Zoological Gardens)

35 Ruffed Lemur *(Varecia variegata).*
The ruffed lemurs' loud call which can swell
into an eerie howling also helps to mark
the territory of these animals that live in
troops. (Cologne Zoological Gardens)

Mongoose Lemur *(Lemur mongoz)*. With a body length of about 35 cm—the tail is equally long—this lemur weighs 2 kilos. Both sexes mark their territory frequently with their anal glands. The baby mongoose lemur is born after gestation lasting from 126 to 128 days, weighs 60 grams, is grayish brown and is carried ventrally by the mother. When 10 days old it begins to clamber about its mother uttering a soft humming sound. It leaves its mother's body for the first time when it is 25 days old, and also begins to develop an appetite for fruit and leaves around this time. These lemurs live in groups of from five to six individuals in the dry forests in the north and northwest of Madagascar.

Crowned Lemur *(Lemur coronatus)*. The color of the male's fur differs greatly from that of the female so it is not surprising that they were previously considered to belong to separate species. The male's fur is reddish and lighter on the underside. The face is whitish, the eyes encircled by a black ring. A broad black patch on the head extends nearly to the ears. The upper side of the female's body is gray, its underside silvery gray to white. The brow is adorned by a golden red triangle, the crown. These lemurs live in family groups numbering from four to eight animals. Gestation lasts from 114 to 128 days and one or two young are born at a time, each weighing from 55 to 60 grams, body length: 10 cm, tail length: 14 cm. When they are a fortnight old the young animals clamber on to the mother's back; at four weeks they go short distances away from her. They start to take solids when they are 30 days old but are suckled for five months. Both sexes mark their territory intensively with their anal glands and urine. This species is rarely seen in zoos. The habitats of these lemurs, whose survival is severely threatened, are the dry forests and savannahs as well as certain zones with dense thickets in the north of Madagascar.

Ring-tailed Lemur *(Lemur catta)*. With its long furry tail ringed with black and white bands, which it uses as a balancing and steering mechanism, this is one of the most conspicuous lemur species. Both sexes have dove gray fur on the upper side of their bodies, a white face, whitish ears and underside, black rings round the eyes and a black mouth. Body length: from 30 to 45 cm, tail length: from 37 to 56 cm, weight: about 2.1 kilos. The head is relatively small. The very mobile ring-tailed lemur is active during the day and lives in groups, spending between 15 and 20 percent of its time on the ground. The females dominate over the males and young animals. They are the first to defend their territory, the males generally remaining in the background. Whereas the females seem to spend all their lives in the same group, the males join other groups. The way they mark their territory is complicated and plays an inportant part in their social life. They use the scent of their hairy armpit glands which they open by pressing them with the horny spurs above their wrists, then rubbing the lower part of their arms against the bark that is thus impregnated with their musk. Only the male has armpit glands. The glands above the wrist of the male are used to rub scent on its tail. Both males and females also mark their territory with the glands on their genitals. The striking fur on the face and tail has also optical significance. When preparing for battle the ring-tailed lemur lifts up its tail and thrashes the end of it toward its head. The effect is visually heightened through a retraction of the upper lip, baring the long canine teeth. The outcome of fights between the males decides their social status within the group. Ring-tailed lemurs bark when excited, purr when content, and before they go to sleep they utter loud owl-like cries which serve to establish the distance between the herds and also contact within the group.

Ring-tailed lemurs eat prickly pears, wild figs, bananas, blossoms, leaves, bark, grasses and herbs. Gestation lasts from 132 to 134 days. The newly born (twins are fairly common) are carried ventrally by the mother. Weighing from 50 to 80 grams at birth, the young have blue eyes to begin with and the iris only becomes lemon yellow later on. The animals are sexually mature when they are two and a half years old.

The ring-tailed lemur's range includes the stony regions and barren mountains in southwestern Madagascar. Since these zones have a dense human population, this species, too, is today endangered. Ring-tailed lemurs have been known to live for more than 20 years in zoos.

Red-bellied Lemur *(Lemur rubriventer)*. The male, which is almost uniformly chestnut brown in color, has a coppery red underside, hands and feet. There is a reddish brown stripe over the brow. The female, umber on the dorsal side, has a cream-colored underside and yellowish white cheeks while the face and forehead are dark brown. They live in groups of four to five animals in the higher regions of Madagascar's eastern rain forests. Little has been discovered so far about these rare prosimians' habits.

Gentle Lemurs (genus *Hapalemur*)

Smaller than the typical lemurs, there is less difference between the length of the arms and legs. They have two pairs of mammary glands, a round head, a short snout and little furry ears hidden in their coat. Active in the day, these prosimians, called "bokumbulo" in Madagascar, are vegetarians, living chiefly on sugar cane and grasses, and in some regions exclusively on bamboo shoots and leaves. They gnaw the hard bamboo stalk with the aid of their sharp incisors. There are two species.

Gray Gentle Lemur *(Hapalemur griseus)*. The body length is about 30 cm, weight 1 kilo. Its sense of smell is very important for orientation. Its marking behavior is interesting. For this purpose it uses a secretion of a milky color and smelling slightly of beeswax from the two armpit glands. Like the ring-tailed lemur, it rubs this scent on to the horny spurs on its wrists and marks various objects in its environment with them. Its threatening behavior is also similar to that of the ring-tailed lemur. It lays its ears back, crouches on its haunches and draws its tail several times between the wrist spurs so that it is impregnated with their scent. The procedure is accompanied by a thrashing of its tail and the uttering of squeals. This species has a very wide range of sounds—squeals, miaows, purring (when content), grunts (for mutual contact), and a loud humming (a call denoting the marking of territory).

They live in groups of from three to five animals. One offspring is born at a time. At birth it is already fully furred, has a body length of 11.5 cm, a tail 10 cm long and it weighs 32 grams. The mother carries it both on her back and in her mouth. At times she deposits it in a nest hole. This species is chiefly found in the bamboo thickets on Madagascar's east coast. It is uncommon in zoological gardens. The more modern classification distinguishes three subspecies:

Hapalemur griseus griseus; Hapalemur griseus alaotrensis; and *Hapalemur griseus occidentalis.*

Broad-nosed Gentle Lemur *(Hapalemur simus).* This very rare species from the reed-covered lowlands on the east coast of Madagascar is indubitably one of the least known members of the lemur family. Larger than the gray gentle lemur it resembles the latter in its fur coloring. The bushy tail is more than half the total length of the animal (90 cm). It has large ear tufts; there are no glands above its wrists and on the upper arms. It mainly lives on bamboo shoots and sugar cane. It eats bamboo by holding it in its hands and peeling off the outer leaves, layer for layer, with its teeth, usually only consuming the fleshy inner sheath. It is rarely seen in zoos and no reports exist of successful breeding in captivity.

Ruffed Lemurs (genus *Varecia*)

According to Petter, they do not belong to the lemur genus but form a genus of their own—*Varecia.* This special status is based on their different behavior patterns, etc. Notable features are shorter gestation, the number of young born at a time, and the fact that they spend their early lives in a nest.

Ruffed Lemur *(Varecia variegata).* Numbered today among the endangered species, the ruffed lemur lives in a relatively restricted area of the rain forests of northeastern Madagascar. It has adapted itself to heavy rainfall by having a thick coat that prevents it getting soaked through. The ruffed lemur is 60 cm long, its tail is of an equal length. It weighs about 3 kilos. The females are usually bigger and heavier than the males. Like the ring-tails, they are great "sun-worshippers." Formerly the local people, believing that the "varikandanas" were sacred animals who worshipped the sun, never harmed them. Nowadays, unfortunately, they are shot.

The female has three pairs of mammary glands. Gestation lasts from 99 to 103 days. Between one and three young are born at a time and the mother deposits them in a nest of leaves and twigs made by her shortly beforehand. Each weighing about 100 grams, the young have light blue eyes at birth that begin to change color after 12 days. To begin with, the mother spends a great deal of her time with the young in the nest to warm and suckle them. After suckling she always licks her babies. The young develop extremely quickly and this is very important since they are not carried around by their mother like the other lemurs. When six days old, they begin to clamber about the nest, at 11 days they start to explore its vicinity, and at 24 days they take the first solids.

The wide range of fur coloring displayed by the ruffed lemurs has given rise to much discussion. The following subspecies have been established:

Varecia variegata variegata. Black and white fur, face black with prolific white side-whiskers and lemon-yellow eyes, black tail. Very attractive. North coast of Madagascar.

Varecia variegata ruber. Face, hands, feet and tail are black; ruff, body, arms and legs fox red. Beige patch on the neck. Occurs in northeastern Madagascar.

Weasel or **Sportive Lemurs** (genus *Lepilemur*)

The coloring of this smaller genus varies a good deal. The dorsal side is gray with reddish tinges, the ventral side lighter. The eyes are small, the ears naked, the hands and feet are long. Whereas all the other lemurs have 36 teeth, these species only possess 32 since they have no upper incisors. The largest unit consists of the female with her offspring; the male lives a few trees further away. After gestation lasting about 135 days one young is born that is suckled for four months and remains with its mother for more than a year. Weasel lemurs mainly live on leaves, but according to season also eat blossoms, fruit, and bark. Because of this purely vegetarian diet their flesh is greatly prized by the local people. The weasel lemurs' diet of leaves is fermented in the large gut and—like rabbits—the excreted pellets are then swallowed again. This is an absolute exception among primates.

The nimble, nocturnally active weasel lemur inhabits all the forest zones of Madagascar. As a result of increasing forest clearance its habitat is becoming more and more restricted. Because of their specialized diet weasel lemurs have so far not been kept in zoos. Even in Madagascar they do not live long in captivity.

The two species **lesser weasel lemurs** *(Lepilemur ruficaudatus)* and **greater weasel lemurs** *(Lepilemur mustelinus)* are the best known. Other species are: *Lepilemur dorsalis, Lepilemur rufescens, Lepilemur leucopus, Lepilemur microdon,* and *Lepilemur septentrionalis* with several subspecies.

Indrisoid Lemurs (family Indriidae)

This family, which includes the largest living prosimians, is composed of the sifakas, avahis and indris. Sifakas and indris have long slender limbs, monkey-like round heads with short wide snouts, very large hands and feet and unusually long legs. The big toe has become a huge grasping digit that can be abducted at an angle of 90° and is used to obtain a hold on vertical tree trunks. The other toes are linked by a membrane up to the top joints. The Indriidae eat leaves, blossoms, bark, and fruit and their digestive organs are adapted to this specialized diet. They have 30 teeth with only two instead of three premolars on each side and no lower canine teeth. They are only found on Madagascar and living specimens are very rare in Europe. It is hard to find a substitute diet for these highly specialized leaf-eaters. There are three genera with four species.

Sifakas (genus *Propithecus*)

The genus *Propithecus* comprises the two species *Propithecus diadema* found in eastern Madagascar and *Propithecus verreauxi* in the west of the island. In terms of their coloring both species have engendered distinctive subspecies. These large animals have long tails which, however, are not used as a steering mechanism since they are not very muscular and droop downward. They serve as a kind of signaling device. The sifaka's fur is very thick and silky, the basic color being white or dove gray, while the face is black and the top of the head brown. Sifakas are social animals, active during the day and living in family groups of from four to five animals. They were named after the sneeze-like sound they utter—"shi-fak." With their long and powerful legs they can leap up to ten meters from tree to tree. When they come down to the ground they stand upright and can hop a distance of almost four meters like a kangaroo.

Gestation lasts five months. Until it is six or seven months old the baby sifaka is carried ventrally by its mother. When it is a fortnight old the young animal begins to clamber about its mother in a lively fashion. When it is three months old it starts to take solids. Formerly the two sifaka species were found in

Sifakas (*Propithecus* spec.)

virtually all Madagascar's forest zones, but today they have been exterminated in many parts. These highly specialized vegetarians only lived for a few weeks when attempts were made to keep them in various European zoos. At the Tananarive Zoo (Madagascar), where they were fed on the buds and leaves of the white mulberry *(Morus alba)*, guavas *(Psidium guajava)* and blades and buds of larger grasses, they lived up to seven years.

Verreaux's Sifaka *(Propithecus verreauxi)*. This species, too, with its four known subspecies, varies a great deal in the fur coloring, even individually. With a body length of from 45 to 53 cm and a tail between 48 and 56 cm long, these prosimians weigh between 4 and 5 kilos. The males of all the subspecies have an elongated patch of scent glands in the region of the throat that are used for marking territory. A characteristic feature is the skin fold with long wisps of hair which extends from the upper arm to the trunk, so that when they jump the sifakas look as if they were gliding. They live in groups of from six to ten individuals (several males and several females) in the west and south of Madagascar. There are four subspecies:

Propithecus verreauxi verreauxi. A white coat, with the exception of the crown of the head which is dark reddish to blackish brown; lustrous yellow iris. The head of the newly born is large, the tail and limbs almost hairless. Found in southwestern Madagascar.

Propithecus verreauxi coronatus. Basic color ash gray to ocher yellow, head velvety black, underside a rusty red. A broader snout than the other subspecies. Found on the west coast of Madagascar.

Propithecus verreauxi deckeni. Pure white without markings, sometimes faded gray or yellow tints on the basic coloring. Found on the west coast of Madagascar.

Propithecus verreauxi coquereli. Conspicuously handsome coloring, basic color white, the inner surface of the limbs and underside of the body chestnut brown.

Diademed Sifaka *(Propithecus diadema)*. Larger than Verreaux's sifaka, the diademed sifaka has a tail that is shorter in proportion to its body. This species lives on the east coast of Madagascar where in the forest regions up to the high plateau five mutually isolated subspecies are found:

Propithecus diadema perrieri. Completely black fur. Far north of Madagascar.

Propithecus diadema candidus. Mainly white, but rusty red fur at the base of the tail. Northeastern Madagascar.

Propithecus diadema diadema. Trunk a light silvery gray, arms and legs a brownish orange yellow. It owes its name to the black to grayish crown which stands out against the otherwise white fur on the head. Eastern Madagascar.

Indri *(Indri indri)*

Propithecus diadema edwardsi. Upper side of the body white, underside grayish white, arms, legs and tail black. Eastern Madagascar.

Propithecus diadema holomelas. Upper side of the body jet black with reddish rump patch. Is distributed in southeastern Madagascar.

Woolly Lemurs (genus *Avahi*)

Distinctive features are a short round head, short snout, large eyes, hairy ears hidden in the thick woolly fur to which this animal owes its name, outsize thumbs, and big toes. The fur is grayish brown; the light-colored fur round the eyes produces an owl-like appearance. Little is known about the avahis' habits. Their food includes leaves, blossoms, buds, bark, and fruit. One young is born at a time and is carried ventrally by its mother. It is suckled for six months. Woolly lemurs are found in the wet forests of the northwestern and eastern parts of Madagascar. They formerly populated the whole of the east coast but today their survival is severely threatened.

Avahi (*Avahi laniger*). The only species of this genus is 30 cm long with a tail length of 40 cm and a weight amounting to 1 kilo, making it the smallest member of the Indriidae family. There are two subspecies:

Avahi laniger occidentalis, upper side olive to light grayish brown; *Avahi laniger laniger*, dark brown; somewhat larger than the other subspecies.

Indris (genus *Indri*)

Indri (*Indri indri*). The only species of this genus, it is the biggest living prosimian, weighing 7 kilos with a body length of 70 or 80 cm and a tail measuring from 3 to 5 cm. Its body is marked with a black and white pattern, the face is naked and black, the short ears are almost hidden in the fur, and it has a jutting dog-like snout and extremely long hands and feet. It is the only Madagascar prosimian whose tail is reduced to a stump consisting of from 8 to 14 small vertebrae, whereas all the other lemurs have from 20 to 32 tail vertebrae.

The number of indris on Madagascar has diminished considerably. They are today still found in the northern half of the eastern forest zone but here, too, they are being forced back into the remaining wooded areas. Indris are hard to observe but betray their presence every morning and late afternoon with long drawn-out wailing calls that can be heard for miles and constitute one of the characteristic sounds of the eastern Madagascar rain forest. These choruses have the purpose of defining their territory. Like the siamang and the orang utan, the indri has a large larynx sac that it uses as a resonator to amplify its voice. Because it is the loudest prosimian on Madagascar the local people call it "amboanala," the dog of the forest.

Indris live in small family communities of up to five animals. Gestation lasts 4.5 or 5.5 months and one young is then born which is first of all carried around ventrally by the mother. At birth the indris are almost entirely black, the white patches developing later. At eight months of age, the indri is independent but remains with its mother. There are many legends about the indris on Madagascar. Because the local people believed that they would be transformed into indris after death they were treated for a long time as sacred animals. Other people thought that men and indris were descended from common ancestors.

Unfortunately the number of indris is still on the wane. Their exclusively vegetarian diet makes it impossible for them to adapt themselves to the changing environment. Several years ago a special reserve for Madagascar's indris was established near Perinet. So far it has been impossible to keep them in captivity, even in the Tananarive Zoo where all the plants on which the indri lives are available.

Aye-ayes (family Daubentoniidae)

In this family there is only one genus (*Daubentonia*) with one living species. In addition, remains have been found of a much larger form, *Daubentonia robusta*, which perhaps only became extinct in the course of recent centuries. *Daubentonia* bears no external resemblance to other Madagascar prosimians.

Aye-ayes (genus *Daubentonia*)

Aye-aye (*Daubentonia madagascariensis*). Originally this odd prosimian was classed with the rodents. Like these, adult aye-ayes have in both upper and lower jaws two huge chisel-like gnawing teeth with open roots and continuous growth, no incisors on the side, no canine teeth and a reduced number of molars. The first to identify this apparent rodent as a prosimian was the German zoologist Schreber around 1775. R. Owen, an English naturalist, produced conclusive evidence by examining the milk teeth of young aye-ayes. He found they still had the incisors on the side and upper canine teeth common to the other prosimians. The aye-aye's round head, large ears, long-haired fur and long bushy tail are also similar to those of rodents. Its kinship with the primates is indicated by the shape of its hands and feet. Its long fingers have claw-like nails. The enormously elongated thin third finger ending in a powerful curving claw plays an important part in its hunt for food. "Aye-aye" is the expression of amazement uttered by local people on seeing this animal for the first time when hunters brought two specimens to the French explorer Sonnerat in 1870. The aye-aye's big toes and thumbs have flat nails. Its fur is usually dark brown. The long straggling side-whiskers are black and often white-tipped.

The aye-aye lives alone or in pairs and feeds on beetles, beetle larvae and other insects, sugar cane, bamboo pith, coconuts, and mangoes. At night it begins its hunt for food by tapping the bark of trees with its third finger. If the bark sounds hollow, indicating the presence of larvae, it gnaws a hole in the tree trunk with its powerful front teeth and extracts the larvae with its long bony finger. The aye-aye also uses this finger as a spoon to scoop out fruit or raise water to its mouth, and as a comb as well. During the day it sleeps rolled together into a ball, its head between its feet and wrapped in its long bushy tail. When moving about, the tail is arched. Thanks to the claws on its hands and feet the aye-aye has no difficulty in climbing trees. Its nest construction is also interesting for it displays an ingenuity rarely found amongst mammals. It often chooses trees in the vicinity of coconut plantations for its nest. To build it, about 60 roughly woven twigs are used. Every day fresh twigs are added to the nest that is in current use. One aye-aye often possesses from 2 to 5 nests that are nearly always built at a height of between 10 and 15 meters. Each nest is completely roofed over and has a side-entrance about 15 cm in diameter. Once the aye-aye is in its nest, this entry hole is often stopped up.

The aye-aye gives birth to one young at intervals of probably two or three years. It is born in the nest, deposited there and suckled for possibly one year. The aye-aye has only one pair of mammary glands and these are situated in the groin.

Its habitat is the dense jungle with bamboo thickets in northeastern Madagascar. Together with the forests, the aye-aye, too, has almost disappeared today. *Daubentonia* is at present the most endangered mammal species on Madagascar. Currently only about 50 specimens are thought to be in existence. A special reservation for these rare prosimians was established on the little island of Nosny-Mangabé in 1966. Thirteen animals were brought there in the hope that they would breed and multiply. Aye-ayes have survived for quite a long time in various zoos. At Amsterdam a specimen lived for 23 years. The aye-aye is included in the IUCN's list of animals threatened with extinction.

Aye-aye *(Daubentonia madagascariensis)*

Infraorder Lorisiformes

The lorisform lemurs in Africa and Asia are nocturnally active tree-dwellers. There are two families: lorises (Lorisidae) and bush-babies (Galagidae) with five genera and 11 species and 60 subspecies.

Lorises (family Lorisidae)

The name "loris" is derived from the Dutch word "loeris," meaning a clown. The arboreal lorises in Africa and Asia are nocturnally active and live on insects and fruit. With their large forward-facing eyes these unusual prosimians can see exceptionally well and spot even motionless prey from a considerable distance. In contrast to the galagos, their ears are small and round. The loris moves with slow deliberation, sometimes hanging like a sloth amongst the branches. It is a silent and adept climber but almost incapable of jumping. The tail has become reduced to a stump for it is not needed as a steering and counter-balancing mechanism. The hands and feet have developed into true grasping organs. The thumbs and big toes are opposable and hence provide the ability to grip. Some of the species have virtually no index finger, which is rarely used, and the third finger, too, is sometimes withered. The second toe has the typical prosimian claw, whereas all the other digits have nails. The loris can cling to a branch in a hanging position for hours on end. The *Retis mirabilis* in its vascular system helps to keep the muscles flexed for a long time without any sign of fatigue. Members of the loris family are common in zoological gardens that sometimes also breed them. Their numbers in the wild are at present not threatened. There are four genera of lorises.

Slender Lorises (genus *Loris*)

Of this genus only one species with six subspecies is known.

Slender Loris *(Loris tardigradus)*. In contrast to the other Asian and African lorisform species this animal has a slender, graceful body and spindly extremities. The tail stump is about a centimeter long. It has gray or yellowish gray fur on the dorsal side and is 25 cm long. Its large eyes, encircled by dark rings covering nearly all the face, are only exceeded in size amongst the primates by the tarsier. The eyes of the slender loris are prized by the local people as a cure for optical diseases, and, made up into love potions, are supposed to be an aphrodisiac. When twi-

light falls, these nocturnal prosimians begin their hunt for food which mainly consists of insects, lizards, birds, eggs, blossoms, leaves, young shoots and unripe nuts. The normally slow-moving animals can pounce on insects with amazing speed. They particularly relish the brains of small vertebrates. The slender loris is not at home on the ground. Its hands and feet, which have developed into grasping organs, are not suited to a smooth flat surface. It usually chooses a forked branch to sleep on, tucking its head well down between its haunches and clinging to the branch with its hands and feet.

Apart from the mating season these animals are not very sociable. Gestation lasts about four months; usually one young is born whose eyes are already open and whose coat is thin and silky. The young are dependent for a very long time, being carried around by the mother for over a year. They reach adulthood when they are 18 months old. The slender lorises mark their territory by urinating on their hands and feet, leaving a pungent scent behind them as they move around. They defend their territory with loud screams. The slender loris lives in the tropical forest regions of South India and in the forests of Sri Lanka.

Slow Lorises (genus *Nycticebus*)

This genus has two species with eleven subspecies which vary considerably in their fur coloring and markings.

Slow Loris *(Nycticebus coucang)*. The slow loris lives in the tropical rain forests of Farther India and Indonesia. Its woolly coat would appear to be an adaptation to the animal's exceptionally low metabolism since even in the tropics it could not survive without this protective covering. With a body length of between 32 and 37 cm the slow loris is considerably larger and more robust than the slender loris. A typical feature is its brown or blackish dorsal stripe. Its backbone is longer than that of other primates and has more vertebrae. The nocturnally active slow loris disappears into holes in the trees during the daytime. It hides from the light by tucking its head between its legs and crossing its arms over it. When twilight falls the slow loris begins its hunt for food. With its fine hearing, excellent sense of smell and good sight it soon detects small vertebrates. The slow loris is more of a vegetarian than the slender loris. It is even capable of climbing a tree-trunk backward with its head upside-down, all in slow motion. Both sexes use urine to mark and define their territory. The slow loris is more socially minded than the slender loris. The animals keep in touch with special contact calls. Gestation lasts 186 days and one young is normally born at a time. When her young is only two days old the mother deposits it on some object to which it clings with its comparatively large hands and feet while she goes on her nocturnal excursions. The first solids are taken when the animal is ten days old. It remains with its mother for nine months.

Lesser Slow Loris *(Nycticebus pygmaeus)*. Also lives in Southeast Asia. With a body length of 19 cm and weighing 500 grams it is half the size of *Nycticebus coucang*.

Angwantibos (genus *Arctocebus*)

Within the genus *Arctocebus* there is only one species with two subspecies that are found in West Africa.

Angwantibo or **Golden Potto** *(Arctocebus calabarensis)*. Its appearance closely resembles that of the slender loris. The angwantibo, whose upper side is rusty brown to gray, is smaller and less deliberate than the potto. The eyes and ears are larger, the snout is pointed and the tail barely perceptible. The index finger has withered away to a wart-like swelling, and the third finger has become a nailless stump. As a result of its nocturnal life in the tree-tops little is known about this odd prosimian. Its diet consists largely of insects; more recent field studies have established that it even eats hairy caterpillars which it previously rubs between its hands to remove the hairs. Like the potto, the genital glands of both sexes exude a pungent secretion indicating fear when danger is in the offing. Apart from mating activities there is little contact between the sexes. Gestation lasts about 131 to 136 days; the offspring weighs from 25 to 30 grams at birth, is covered with fur and its eyes are open. It is normally carried around ventrally by its mother. The color of the young animal's fur is much darker than that of the adults. When it is a fortnight old, its mother already deposits it on a twig while she forages for food. Later on the young animal utters a click-like call that indicates its whereabouts to the mother. Young angwantibos are weaned at four months and growth is completed at seven and a half months. These animals are found in a relatively small area of the West African tropical forests. Possibly the most rare species of loris, they are today in need of protective measures. So far the angwantibo has seldom been kept in captivity.

Pottos (genus *Perodicticus*)

Potto *(Perodicticus potto)*. The only species of this genus, with five known subspecies, the potto lives in the forests of East, Central and West Africa, weighs one and a half kilos, has a round head with short snout, medium-sized eyes, fairly small ears and a tail about 6 cm long. It has specialized tactile hairs on its neck that serve the sense of touch. An unusual feature of the nocturnally active potto is that it has long dorsal spines rising

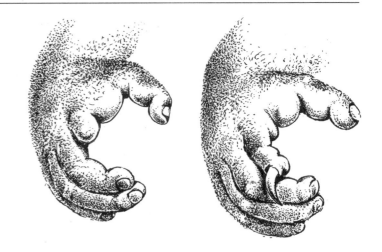

Hand and foot of a potto *(Perodicticus potto)*.
With the index finger reduced to a stump the hand is an even more effective grasping organ.

from its backbone and protruding above the surface. If the potto is attacked it lowers its head to expose the hard prominences to its aggressor. The thumb is very large and is abducted from the other fingers at an angle of 180° and the index finger is reduced to a stump so that the hand forms a kind of forceps with a wide grip. The big toe is abducted in a similar way so that the foot, too, has great grasping power. There is only one claw and that is on the second toe. Eighty percent of the potto's diet is vegetarian, the remainder being of animal origin. Gestation lasts from 6 to 6.5 months and one young (rarely two) is born at a time. Young pottos have a silvery white coat which turns into the yellowish brown fur of the adult animal after six months. During the first weeks of its life the young potto is carried ventrally by the mother. When it is a month old it rides on her back, at two or three months it begins to take solids and is adult at nine months. Pottos are successfully kept and bred in the nocturnal animal houses of modern zoos.

Galagos or Bush-babies (family Galagidae)

The socially living bush-babies are classified as near relations of the lorisform lemurs. The principal difference is that they are more mobile than the lorises. They gained the name of bush-baby on account of their wails that sound like a crying infant. They have long tails, elongated tarsal bones and are phenomenal leapers. Like the tree shrews, their tongues are underlain by a cartilaginous sub-tongue which is used for grooming purposes. The large membraneous ears are pliable and also have a sizeable flap whose cartilaginous tissue and small muscles enable it to be folded over when the animal is asleep. The neck is so flexible that the head can be swiveled through 180°. The eyes are front-facing and have a thick pigment layer behind the retina *(Tapetum lucidum)* which at night reflects the rays of moonlight and bright stars, thus attracting night moths so that the bush-babies only need to make a lightning-like grab. They are omnivorous. They mark their territory, its borders and their nests frequently by urinating on their hands and feet. Only found in Africa and on certain adjacent islands, there is so far no threat to their survival. They are often seen in the nocturnal animal houses of zoological gardens. The present classification of existent galagos according to species and subspecies cannot be considered satisfactory. There is only one genus *(Galago)* with six species and 28 subspecies.

Galagos (genus *Galago*)

Senegal Bush-baby *(Galago senegalensis)*. The species with yellowish brown fur on the upper side of the body represents the true bush-baby. In appearance and anatomy it resembles the great bush-baby, but with a head and body length of from 16 to 20 cm and weighing from 150 to 300 grams is only half the size. In its natural habitat the bush-baby avoids the forest floor. When it comes down to the ground it stands upright like a kangaroo and hops on its hind legs. The male has a naked triangular patch of glands on the middle of its chest and uses the scent from these glands as well as urine to mark its territory. Nocturnally active, they are sociable animals which—except during the mating season—live in supra-family groups (up to nine adults). During the mating season the pairs go off on their own. There are often battles between rival males at this time. They threaten each other by opening their mouths, extending their hands and snarling loudly. The female also avoids the male when her young are due to be born. Gestation lasts from 120 to 146 days; one or two furry young weighing from 8 to 22 grams are born with their eyes open. Young bush-babies have practically no clinging reflex. The mother never carries her young around with her but leaves them alone in the nest during the first two weeks while she makes her nocturnal forays. Apart from gum resin the bush-baby is an insectivore. The Senegal bush-baby and its numerous subspecies are found all over Africa south of the Sahara.

Great or **Thick-tailed Bush-baby** *(Galago crassicaudatus)*. The largest galago is represented by numerous subspecies which occur from Central Africa to the southeastern part of East Africa and by one subspecies in West Africa. They are chiefly found in the dense forests. Head and body measure 33 cm and the woolly fur is yellowish gray on the dorsal side and gray on the ventral side. The bushy furred tail is 37 cm long and the animal weighs between 1 and 1.25 kilos. The base of the tail is used to store fat on which it subsists when sleeping through the dry season. The adult male also has a naked patch of glands under the chin and uses this scent to mark its territory. In addition it sprinkles urine on its hands and feet and then makes pungent imprints with them. An interesting feature is the upright stance when engaged in combat. The animals stand face to face, exchanging blows with their powerful fore limbs. Their diet includes fruit, seeds, nuts, berries, leaves, buds, sap, insects, small birds, and eggs. Living prey is caught with a lightning swoop, bitten to death and chewed. The great bush-baby's enemies are large owls, wild cats, and snakes. This prosimian, too, is relatively sociable; members of a family sleep together in their nests of leaves. In the wet summer months one young is born after gestation lasting from 126 to 136 days. The newborn has a thin coat of fur and with a body length of 10 or 12 cm weighs from 50 to 70 grams. The mother deposits it in a nest and only carries it in her mouth when danger is in the offing. The young make their first forays out of the nest when they are two weeks old. They are suckled for three months. There are 11 subspecies.

Allen's Bush-baby *(Galago alleni)*. Much smaller than the great bush-baby, it has conspicuously large ears. This species, light to reddish brown on the dorsal side and yellowish on the ventral side, weighs about 250 grams and has a body length of from 20 to 22 cm. It is considered the best leaper of the family. It lives in Central West Africa where four subspecies have been identified.

Demidov's Bush-baby *(Galago demidovii)*. We know little about the life of this species in the wild since it is a forest-dweller and therefore much harder to observe than species found in the grasslands. These little animals that only weigh from 50 to 80 grams and whose fur is brown or yellowish brown, live in Central Africa. With their short pointed snouts they resemble the Madagascar mouse lemurs, but their leg joints are greatly elongated, enabling them to make horizontal or vertical leaps of up to four meters. Their ears are shorter than those of the other galago species; the tail is less hairy. In their natural habitat they mainly live on insects, tree sap and fruit. One or two young are born after gestation lasting from 108 to 110 days, and with a head and body length of 7.5 cm and tail measuring 9 cm weigh up to 10 grams. At the age of three weeks the young can already run quickly, after three and a half weeks they take solids, are weaned at six weeks, and are independent at the age of three months. There are seven subspecies.

Needle-clawed Bush-babies are tree-dwellers of the tropical African rain forest zones. Some zoologists regard them as a distinct genus—*Euoticus*. There are two species: the **Western Needle-clawed Bush-baby** *(Galago elegantulus)* and the **Eastern Needle-clawed Bush-baby** *(Galago inustus)*.

The needle-clawed bush-babies have a thick, short-haired, woolly coat and short and strong arms and legs. The long tail of the western needle-clawed bush-baby is very hairy, whereas that of the darker colored eastern species is covered with short hairs. Their naked ears are pink, hands and feet are also hairless and flesh-colored. The eastern needle-clawed bush-baby is smaller, but has larger ears. They owe their name to the shape of the toe and finger nails which, instead of being flat, have a central ridge along their length ending in a needle-fine point. These pointed nails give them a hold on smooth tree trunks along which they can also run head downward. They mainly feed on sap but also eat insects and fruit. The western species has two subspecies.

Infraorder Tarsiiformes

Tarsiers (family Tarsiidae)

Tarsiers (genus *Tarsius*)

The tarsier family occupies a special place amongst the prosimians because it has lost many ancestral features. The last living member of the Tarsiidae is the tarsier, today found in the tropical forests of Southeast Asia. It is seen as the final link in the evolution of the prosimians and a close relative of the true monkeys. Analyses of its blood serum have corroborated this theory. The formation of the outer auditory canal, the round skull with an opening at its base, the reduced size of the nasal passages and the partial separation of the eye sockets from the temporal bone as well as its dentition are all monkey-like in character. The lower incisors are almost vertical, the lower canine teeth do not resemble incisors as in the case of many other prosimians, but are genuine eye-teeth. The dental formula is $\frac{2 \cdot 1 \cdot 3 \cdot 3}{2 \cdot 1 \cdot 3 \cdot 3}$. Fossil remains indicate that tarsiers used to inhabit Eurasia and North America. During the Tertiary period tarsiers evolved into a surprising number of forms. Today only remnants survive of this once so successful group which about 60 million years ago split off from the ancestors of the other primates. It is of course not possible to draw definitive conclusions about the stage of development or degree of kinship from adaptations pointing in the same direction. Unfortunately the extinction of the tarsiers is drawing near, since suitable habitats are rapidly disappearing as a result of forest clearance and cultivation. Amongst the local people of the Malay archipelago the nocturnally active tarsiers are regarded as unearthly creatures and harbingers of misfortune. Tarsiers are small and highly specialized primates. Their average total length amounts to 40 cm, with the near-naked tail, tufted at the end, measuring about 24 cm. The bulging round eyes are enormous and occupy most of the face. The pupils are so much enlarged at night that the iris is only visible as a tiny ring. Over the eyes there are conspicuous stiff sensory bristles. The large round head—that can be swiveled through 180°—has a short snout and pliable ears that can each be moved separately and are constantly in motion when the animal is agitated. The tarsier relies much more on its hearing than on its sense of smell. Its tarsal bones are much elongated (a feature to which it owes its name) with the result that it has become an excellent leaper. The skeleton is adapted to rapid leaping through the partial fusion of the tibia and fibula and the elongation of the heel region. The best leaper amongst the primates, it hops with its hind legs and can leap one meter. Just before landing it extends its limbs and, using the broad flat

Tarsier *(Tarsius)* jumping.

pads on its fingers as a suction mechanism, obtains a firm hold on the branches. The tail acts as a support in grasping thin twigs. These primates prefer lizards and insects to other fare. When grabbing its prey the tarsier shuts its eyes to avoid injury to them. It catches insects on the wing by leaping at them. It also angles small fish and crabs out of streams. The tarsier is probably the only primate that does not eat any plants. It lives alone or in monogamous families in territories that it constantly marks with urine or a secretion from its anal glands. Gestation lasts six months—an exceptionally long time for such a small animal. So far no evidence has been produced to indicate that the tarsier builds a nest. Despite the two pairs of mammary glands only one young seems to be born at a time. At birth it is furry and its eyes are open. The newly born clings to its mother's underside with hands and feet.

The genus *Tarsius* consists of three very similar species. Occasionally taxonomists only classify one species. The primatologists divide each species into several subspecies.

Philippine Tarsier *(Tarsius syrichta)*. This is the largest species. With a body length averaging between 8.5 and 16 cm and a tail from 13.5 to 27 cm long, the grayish red brown primate weighs from 85 to 165 grams. Characteristic features are the naked smooth under-surface of the tail and the hairless feet and elongated skull. It was the first tarsier to be described and depicted in 1702.

Western Tarsier *(Tarsius bancanus)*. Its head is shorter and the fur is a grayish golden brown color. It mostly occurs in coastal forests and in the vicinity of rivers on Kalimantan and southeastern Sumatera.

Eastern Tarsier *(Tarsius spectrum)*. It has hairy feet, a white patch behind the ears and the under-surface of the tail is covered with bristles. It is found on Sulawesi.

Monkeys and Apes (suborder Simiae)

In terms of their evolutionary history the monkeys and apes are the most successful primates and differ significantly from all the prosimians. Napier and Napier (1967) call this suborder Anthropoidea. In some regions they are found in a bewildering number of species and almost indistinguishable subspecies. One reason for this is that primate populations were isolated from each other by the widening of rivers or the division of originally large belts of forest. In the course of many generations these populations underwent variations, mutations or genetic recombinations so that where favorable selective influences prevailed new and mutually differing forms evolved.

Taken as a whole monkeys and apes present a very variegated picture, ranging from the squirrel-size species to the powerful heavyweight gorillas. Nowadays there is no longer any doubt about the fact that great apes and human beings have more in common than, for instance, the gorilla and baboon. Generally speaking, there are two groups of anatomical features which in various modifications can be identified as common to the New and Old World monkeys and the great apes. These include the generally roundish head with front-facing eyes providing binocular sight, the humanoid form of the ears and the more or less human-like cast of the frequently bewhiskered face that otherwise is sparsely haired. The more highly developed facial musculature permits an expressive play of features. With from 32 to 36 teeth, the dental formula is $\frac{2\cdot1\cdot2-3\cdot2\cdot3}{2\cdot1\cdot2-3\cdot2\cdot3}$—a set of teeth capable of coping with an omnivorous diet. The other common features are the hands and feet which have a fairly true grasp ("four-handers"); the thumbs and big toes are opposable to the other digits.

All simian primates have two mammary glands on their chests and genuine sweat glands. Only a few species form exceptions to this rule—for instance the marmosets and tamarins with their claw-like nails or the baboons with their long dog-like muzzles and formidable fang-like canine teeth. Some groups have almost completely lost their tails (great apes), others still possess a stump (some of the macaques), while a large group possesses remarkably long tails (guenons, langurs).

Simian primates can mate at all seasons of the year. In most cases an ovum is produced at four-week intervals. At this time of increased hormone-controlled sexual receptiveness, the outer genitalia of many female Old World primates swell and redden to a varying degree. This is a kind of signal that triggers off the mating urge. A similar signaling device amongst some groups is the startling bright color of the scrotum. The penis hangs limply from the lower abdomen and erection occurs when mating takes place. In contrast to the prosimians the uterus is a compact pear-shaped organ.

All simian primates live in social groups in the tropics; only a small proportion is found in sub-tropical and more temperate mountain zones. They are mainly herbivores but some of them also eat food of animal origin.

Even more important than the previously mentioned features is the more highly developed brain and nervous system. Amongst the monkeys there are indeed lower forms with a simple brain structure and practically no convolution of the cerebral cortex, but the more highly developed species have brains with much better association centers where the capacity to learn and memorize is located. This evolutionary advance, together with the previously mentioned human-like anatomical features, the trend towards bipedalism, leaving the hands free for improved manipulation, was the point of departure for the development of the human race.

In conventional taxonomy the simian primates are divided into the infraorders: New World or flat-nosed monkeys (Platyrrhina) and Old World or thin-nosed simian primates (Catarrhina).

New World or Flat-nosed Monkeys (infraorder Platyrrhina)

All monkeys are probably descended from a complex group of prosimians that about 50 or 60 million years ago lived in North America and Europe which formed a single continent at that time. At a very early stage one branch reached Asia and Africa and another South America which was still connected to the northern part of the continent by an isthmus. When this bridge disappeared in the early Tertiary period as a result of geological changes the monkeys in South America became isolated and developed along their own lines for about 30 million years. Although both Old and New World primates, as a result of their largely parallel evolution have much in common both physically and in their habits, the latter have retained their ancestral anatomical features. As the scientific name indicates, the New World monkeys have flat noses with widely spaced nostrils opening to the side. The South American monkeys have great difficulty in opposing their thumbs to the other digits or cannot do this at all; that is to say their thumbs are not true grasping organs. On the other hand, their feet have abducted grasping toes like their Old World relatives. They have three premolars in each of their jaws (an exception being marmosets and tamarins).

The majority of flat-nosed monkeys still have brains with a simpler structure. They have no hard patches (ischial callosities) on their hindquarters. A highly specialized feature is the prehensile tail which is tantamount to a fifth limb and is predominantly found amongst the spider, woolly and howler monkeys.

Sound is of supreme importance as a means of communication although the skin glands and the corresponding sense of smell also play a major role in this respect. No New World monkeys live on the ground. The South American monkeys nearly all inhabit the extensive jungle zones. The long broad rivers form impassable barriers so that often members of a genus or species living on either side of a river differ greatly from each other in appearance.

The infraorder of New World monkeys is divided into three families: Cebidae, Callimiconidae and Callithricidae.

Family Cebidae

Both in appearance and habits this group covers a wide range. Their size varies from that of a squirrel up to that of a large domestic cat. They have 36 teeth. There are five subfamilies with a total of 11 genera and 35 species.

Night and Titi Monkeys (subfamily Aotinae)

Night Monkeys (genus *Aotes*)

Douracouli *(Aotes trivirgatus)*. The genus *Aotes* only possesses one species. Some primatologists regard the douracouli, despite its specific adaptation, as the most ancestral monkey. They are the only monkeys that are nocturnally active, at their busiest after sunset and before daybreak. Head and body measure from about 30 to 36 cm, the hairy tail is 50 cm long; their weight varies between 0.8 and 1.1 kilos. The douracouli has large nocturnal eyes with a bulging cornea surrounded by rings of white hair. Its thick coat is gray, beige or dark brown in color. It has tactile pads on the tips of the fingers. In pairs or small family groups the douracoulis prowl sure-footedly and noiselessly through the canopy of branches and are also adept jumpers. Their diet consists of fruit containing water, sugar and oil, seeds, fresh leaves, sap, buds as well as a fair quantity of large insects, small birds, spiders, and tree snails. The pairs sleep during the day, always sharing the same holes in trees.

The douracoulis inhabit relatively small areas of the South American tropical forest where they find sufficient sustenance the year round. This habitat extends from northern Colombia, Venezuela, Peru, northeastern Brazil as far as the Gran Chaco. They mainly mark their territory with urine. Like most of the New World monkeys they are furnished with scent glands so that olfactory communication plays a more important role than in the case of Old World monkeys. Their behavior is naturally adapted to their nocturnal way of life. M. Moynihan (1976) observed, for example, that mutual communication through ritualized behavior patterns such as threatening postures were restricted to gestures performed with the entire body. The more subtle means of expression that can only be registered by sight, such as threatening grimaces or bristling fur (piloerection), would in any case not be perceptible in the darkness. For the douracoulis grooming is a preliminary to sexual activity.

Gestation lasts about 150 (?) days and twins are usually born, each weighing from 90 to 100 grams. During the first weeks of their lives they cling to the fur on the backs of the mother or father. They are reputed to have a maximum live expectancy of 25 years. Douracoulis are only occasionally found in zoological gardens.

Titi Monkeys (genus *Callicebus*)

In size and appearance the titis resemble the douracoulis. They are generally highly colored and have long hair on the sides of their heads. When at rest their typical posture is to squat across a branch, the relatively long tail drooping limply and the weight

Titi monkeys (*Callicebus* spec.) at rest.
The close bonds between the pairs are also expressed by the affectionate way they sit together with entwined tails.

of the body being borne by the feet. When movement is called for this monkey can suddenly extend its legs to take a flying leap, either to swoop upon prey or to seek refuge in flight. Like the marmosets and tamarins their nails are elongated. Little is known about their habits in the wild. U. Hick describes them as being amicable animals. A population density of up to 500 individuals per square kilometer has been registered in certain places. They probably live in pairs, usually with two young of different ages. Gray titis are also found in somewhat larger groups. A remarkable feature is the way they maintain contact with their tails, for instance by loosely entwining them, rubbing them together or through light mutual pressure. Scent from their chest glands probably serves the marking of territory. They communicate with their resounding voices, breaking into a loud chorus particularly in the morning sunshine. Threatening behavior takes the form of bared teeth, rolling eyes and bristling fur. Their diet consists mainly of small animals, insects, lizards, small birds, eggs, etc., but sometimes also soft, juicy fruit, leaves, and blossoms. Occasionally they even use stems to extract insects from crevices (Bungartz). In captivity these monkeys call for a great deal of care so they are not common in zoos.

There are four species with 24 subspecies. Since it is difficult to differentiate between the various forms and to establish the limits of their distribution with certainty, only the species are named here:

Collared Titi Monkey *(Callicebus torquatus)*. Body length: up to 45 cm; length of tail: 50 cm. Coloring: dark brown, chest reddish brown, arms and legs black, a white bib, face black. The coloring varies locally. Range: the northern part of South America from eastern Colombia over Venezuela to French Guiana.

Red Titi Monkey *(Callicebus cupreus)*. Not so brightly colored. All subspecies have mainly reddish fur on their undersides. They have a disjunct distribution in the central and upper Amazon regions as well as in northwestern Colombia.

Orabussu Titi Monkey *(Callicebus moloch)*. The best known species. Reddish brown in color with gray outer surface of the arms and crest, whitish yellow ruff. The subspecies include Hoffmann's Titi *(Callicebus moloch hoffmannsi)* with a whitish ocher ruff. Found in northwestern Brazil and at the mouth of the Amazon.

Masked Titi Monkey *(Callicebus personatus)*. Has predominantly dark to black fur on its chest and front of the head. Found in southeastern Brazil.

Saki Monkeys (subfamily Pithecinae)

With a few exceptions, this group, too, inhabits the impenetrable Amazon virgin forests. Slender in build, they usually have a coat of long thick fur that is slightly greasy to the touch and protects against rain. As grasping organs the thumbs and index fingers can be opposed to the other digits. Three very unusual genera with nine species are summarized here.

Saki Monkeys (genus *Pithecia*)

Their most conspicuous features are very long and coarse fur and a long bushy tail. The larynx with its enlarged frontal cartilage is an effective resonator. Their weight can amount to about 1.8 kilos. They jump with agility over wide and deep puddles or travel hand over hand from tree to tree. Sometimes they also run on their feet along stout branches. Like the gibbons, they raise their arms to help keep their balance, but only sideward. In order to have their hands free for feeding they sometimes hang downward, clinging to a branch by their feet. Both in small family parties or in larger groups they travel through the forest where they find an abundance of animal and vegetable food. In order to drink they plunge their hairy hands into the water and then suck off the moisture. They communicate with certain gestures, facial expressions, postures and sounds.

In captivity they are difficult charges, so are rarely seen in zoos. There are two species with several subspecies.

Hairy Saki *(Pithecia monacha)*. Body length: 45 cm; tail length: 40 cm. Coloring: dark fur tinged with gray especially on the head and shoulders; the coarse and shaggy fur which gives the animal its name often looks like a wig. Range: Central Amazon Basin (Brazil), Central Colombia, Guayana, Surinam.

In 1978 there were only nine hairy sakis in six zoological collections. One of these monkeys lived for more than seven years at the Philadelphia Zoo.

White-faced Saki *(Pithecia pithecia)*. Body length: about 37 cm; tail length: 36 cm; somewhat smaller than the previous species. The most noticeable feature is the male's light-colored coarse-haired mask which surrounds the face with the exception of the black, almost hairless, nose, mouth and eyes. The body is black. The female can be identified by a light stripe over the cheeks on each side of the nose; the blackish fur is white tipped. Range: southern Venezuela. At the Cologne Zoological Gardens young of this species have often been born and it was observed that gestation lasted from 163 to 176 days. After 16 days the baby was seen to be carried on its mother's back and at the age of four weeks on that of the father. There are two subspecies:

Pithecia pithecia pithecia: the male has a white or yellowish white face mask; *Pithecia pithecia chrysocephala:* the male has an ocher-yellow face mask.

Bearded Sakis (genus *Chiropotes*)

The body measurements are similar to those of the other sakis. The round bushy tail is longer than the body. Certain features of the skull differ from those of other species. A very remarkable feature is the hair on the head of some varieties. At the crown the hair forms a crest which sometimes hangs down at the sides corkscrew fashion. In fact their fur looks very "well-groomed." Here, too, a wide range of typical coloring is found. These monkeys live on fruit, preferably berries, blossoms, insects and frequently small mammals and birds. Their habits are quite unknown. Occasionally the species are classified as subspecies.

Black Saki *(Chiropotes satanas).* Its body coloring varies between chestnut, dark and blackish brown. Adult males possess magnificent side-whiskers and the crest is parted in the middle. This monkey occurs in the wet hot tropical forests south of the lower course and mouth of the Amazon.

The black saki is very rarely seen in zoos. In 1960 there was a 15-year-old black saki in San Diego Zoo.

Red-backed Saki *(Chiropotes chiropotes).* The males of this species may grow into imposing creatures. The shoulders and upper part of the back are dark brown, the upper arms chestnut brown. The hair on the cheeks, chin and head grows to astounding lengths. Like the black saki, the scrotum has very distinct naked patches. The long thick fur gives the tail a bushy appearance. When agitated the animal usually stands erect, bares its teeth and lets its beard quiver. This species is found in the tropical forest valleys between the lower course of the Amazon and the Orinoco. Specimens are very uncommon in zoos.

White-nosed Saki *(Chiropotes albinasa).* This rare species has a glossy black coat with a thickly haired tail. The nose and mouth-parts are pink and covered with sparse white hairs. The large, gentle and eloquent eyes are of great beauty. Its area of distribution in Brazil is south of the Amazon between the Xingú and Tapajos rivers. In 1978 only four specimens were living in captivity. Classified by the IUCN as rare species.

Uakaris (genus *Cacajao*)

The uakaris are the only New World monkeys with short tails. The head and body length is between 40 and 45 cm, while the tail is only 20 cm long. They live almost entirely in the higher arboreal regions. If they come down to the ground they prefer to walk upright, keeping their balance by raising their arms. The body appears more thickset than that of the sakis. The main diet consists of a medley of fruit, green shoots, and blossoms found in abundance in the wet jungle. They also eat small lizards, amphibians, birds and insects. The local people occasionally keep tame uakaris as domestic pets. The animals are captured with the aid of arrows smeared with diluted curare. As soon as the poison begins to take effect and the animal collapses, it is given salt which is considered to be an antidote. Classified into four species, all of which are regarded as rare, the uakaris call for great care in captivity.

Red Uakari *(Cacajao rubicundus).* The slender body of this most common and well known species has a thick coat of red to reddish brown fur. The relatively thin extremities have large hands and feet. The face, sides of the head and front part of the crown are bald; forehead, nose and chin are sparsely haired. The red fur on the head is short. A notable feature is the unusual length of the canine teeth. In their jungle habitat these animals are very lively, climbing, leaping and moving hand over hand, running along twigs and also clinging by their outstretched arms to branches. They threaten by opening their mouths wide. When agitated, the face turns even redder, and is paler when the animal is unwell.

Their area of distribution is in northwesterrn Brazil between the Río Negro and the Japurá and they are said to live in families and also in groups. The first birth in Europe took place in Frankfurt on Main Zoo in 1967. They have been known to live in captivity for over ten years. The red uakari is also classified as a subspecies of the scarlet-faced uakari *(Cacajao calvus rubicundus).*

Scarlet-faced or Bald Uakari *(Cacajao calvus).* Similar to the previous species. Its red face and bald forehead form a distinct contrast to the whitish fur. When the animal is agitated, the red color deepens. The males are agile, very mobile, powerful and can turn aggressive. This species occurs in a relatively small zone between the Japurá and the southern part of the Ica river. Very rarely seen in zoos, these animals are registered as an endangered species.

---

36 Douracouli *(Aotes trivirgatus).*
The visibly large larynx sac acts as a resonator. Especially in the early morning the douracoulis howl in chorus making an almost incredible noise for such small creatures. They have a repertory of almost 50 sounds for mutual communication within the group. (Duisburg Zoological Gardens)

Previous pages:

37 White-faced Saki *(Pithecia pithecia pithecia)*.
Only the male animal has a mask of coarse
light-colored fur on its face. (Cologne
Zoological Gardens)

38 Hairy Saki *(Pithecia monacha)*.
This monkey owes its name to its loose-lying
and shaggy coat. (Cologne Zoological Gardens)

39 Hairy Saki *(Pithecia monacha)*
(Cologne Zoological Gardens)

40 White-faced Saki *(Pithecia pithecia pithecia)*.
The Cologne Zoological Gardens has had
repeated breeding successes with this little
monkey. (Cologne Zoological Gardens)

41 Hairy Saki (*Pithecia monacha*).
Only a few zoological gardens possess specimens of this monkey. (Cologne Zoological Gardens)

Overleaf:

42/43 Scarlet-faced or **Bald Uakaris** (*Cacajao calvus*).
These rather silent monkeys seem to wear a sad expression. (Cologne Zoological Gardens)

44–46 Red-backed Sakis (*Chiropotes chiropotes*).
Above: The hair on the top of the head
forms a crest that falls in all directions.
(Duisburg Zoological Gardens)
Above right: All saki monkeys have an
exceptionally thick woolly coat that protects
these animals, which live in the tree-tops,
from the rain. (Cologne Zoological Gardens)
Below right:
The long dark beard is a very conspicuous
feature. (Cologne Zoological Gardens)

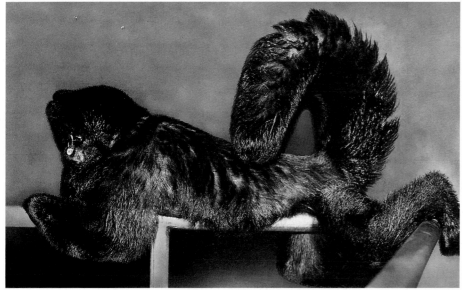

47 Red-backed Saki *(Chiropotes chiropotes)*.
The rounded bushy tail droops slightly at
the tip. (Cologne Zoological Gardens)

48/49 White-nosed Sakis *(Chiropotes albinasa)*.
This rare species is only to be found in
a few zoological gardens. (Cologne Zoological
Gardens)

Overleaf:

50 Goeldi's Monkey *(Callimico goeldii)*.
Little has so far been discovered about the
wild-life habits of this animal which is not so
very uncommon in its homeland.
(Cologne Zoological Gardens)

51 White Tamarin *(Saguinus melanoleucus)*.
A rarity in zoological gardens. (Cologne Zoological Gardens)

52 Woolly Monkey *(Lagothrix lagothricha).*
Woolly monkeys are peaceable animals and in
captivity make friends with human beings
and other animals too. (Liberec Zoological
Gardens).

53 Squirrel Monkey *(Saimiri sciureus).*
The little round head, high domed forehead,
large round eyes and short muzzle give even
adult animals a juvenile appearance.
(Děčín Zoological Gardens)

54 Woolly Monkey *(Lagothrix lagothricha).*
Woolly monkeys use their arms, legs and tails
alternately both for normal locomotion and
also when fleeing. (Liberec Zoological
Gardens).

55/56 Common Marmoset
(Callithrix jacchus).
This marmoset has on each ear an upstanding tuft of long stiff white hair which forms a distinctive contrast to the rest of its coat.
Above:
Liberec Zoological Gardens
Below:
Stuttgart Zoological Gardens

57 Black Spider Monkey
(Ateles paniscus).
The crest of "windswept" hair over the forehead is a typical feature of the spider monkeys.
(Dresden Zoological Gardens)

58 Spider Monkey *(Ateles spec.).*
The long thin prehensile tail is very flexible and its owner can hang by it for hours on end.
(Hanover Zoological Gardens)

59/60 Imperial Tamarins (*Saguinus imperator*).
The enormously long moustache almost touches
the arms. Originally only dead specimens were
available, and they were erroneously stuffed
with the points of the moustache turned upward
in the manner of the German Kaiser Wilhelm
with the result that the species was given
this imperial name. (Cologne Zoological Gardens)

61 Cotton-head Tamarins
(*Oedipomidas oedipus*).
The large white mane stands on end
when the animal is agitated or in display. (Cologne Zoological Gardens)

62 *Saguinus bicolor bicolor.*
This pied tamarin subspecies has a naked black or brownish head with large jug-handle ears giving it a strange appearance. (Stuttgart Zoological Gardens)

63 Cotton-head Tamarin *(Oedipomidas oedipus).* Nowadays these animals are often used for experimental purposes in behavioral research. (Cologne Zoological Gardens)

64 Golden Lion Marmoset *(Leontideus rosalia).* Lion marmosets used to be fairly common in European zoological gardens but are today very rarely seen. (Prague Zoological Gardens)

Black-headed Uakari *(Cacajao melanocephalus)*. As the name indicates, this animal's face is black. Its body fur is also mainly black, the rest being chestnut brown. Its range is the wide valley of the Río Branco in northern Brazil and parts of southern Venezuela.

In 1800 Alexander von Humboldt discovered this species, the first uakaris known to zoologists. Since then regrettably few details about its habits have been established. In 1978 only three specimens were known to be living in captivity. The wild-life number of this species is uncertain.

Black Uakari *(Cacajao roosevelti)*. Not all zoologists regard this animal as belonging to a separate species. It is possibly just a subspecies of the black-headed uakari. With a body length of 45 cm it is considered to be the largest uakari; the tail, measuring about 39 cm, is relatively long. Like the black-headed uakari, its face and fur are black. It is frequently to be found in a disjunct area in the eastern parts of Bolivia. Little is known about its habits there.

Capuchin Monkeys (subfamily Cebinae)

This subfamily has two genera—capuchins and squirrel monkeys. Up to 25 years ago there was a clash of opinion about their correct classification. I. T. Sanderson and G. Steinbacher, for instance, placed the squirrel monkeys next to the sakis, probably because of certain superficial resemblances. Major taxonomic criteria, however, justify their being grouped together with the capuchins, the "true" or "typical" New World monkeys.

Squirrel Monkeys (genus *Saimiri*)

The body length of these little monkeys averages between 27 and 35 cm, the tail is about 40 cm long and they weigh from 800 to 1,200 grams. Occasionally much larger specimens are found. The slender, short-haired body is mainly yellowish, light brown to reddish in color and there are whitish markings round the eyes, ears and throat as well as on the sides of the neck. The nose and mouth look as if they had been dipped in ink or blueberry juice. Another rather unusual feature is the shape of the monkey's skull. The bony capsule holding the brain is elongated toward the back. The brain itself is very large in proportion to the animal's weight, thus it has the relatively largest brain of all non-human primates. The motor areas are the most highly developed.

Gestation lasts about 180 days and a large baby is then born whose weight at birth is about a sixth or an eighth of that of the mother (1/19th for humans). The baby clings tenaciously to the mother who hardly pays any attention to it, at least during the first days, unless it needs help. After three weeks other females begin to take an interest in the baby, letting it ride on them. The squirrel monkey's diet is composed of insects with the addition of snails and tree frogs, as well as fruit and other vegetable substances.

Squirrel monkeys show a marked preference for impenetrable parts of virgin forest. As genuine herd animals—according to R. W. Thorington (1968) they live in groups numbering from 30 to 150 individuals and in Guayana I. T. Sanderson observed a group of over 500 individuals. They hunt for food almost noiselessly among the twigs in smaller groups numbering from five to ten animals but always keeping within sight of each other. In regions with dense tree cover two or even three groups often remain within sighting distance. These monkeys rarely come down to the ground. By impending danger the whole herd retreats screaming to a distant thicket, remaining silent there until they feel sure they are in safety. Their enemies are the larger birds of prey, snakes and, more rarely, medium-sized jungle cats. Like some of the guenons they utter specific warning calls to alarm other members of their group in cases of danger.

Their social behavior is well developed. They spend more than half the day watching each other and engaging in social communication. The hierarchy, however, has been found to be instable. The males often live in a state of semi-independence from the female groups. In their mating behavior olfactory communication plays a prominent role. R. Kirchshofer wrote that the animals mark or "impregnate" themselves with their urine. They are nevertheless very clean creatures. Their sexual behavior is interesting because it takes the form of ritualized threat and dominance behavior. The male stands upright, stretching one leg away from the body so that the erected penis is clearly visible—a display posture that can be observed even amongst very young animals. The habits of these monkeys have been fairly well studied. Large numbers of them are still supplied to testing and research laboratories.

In captivity they require spacious accommodation with enough room to move about in. Their lack of resistance to parasites means that they must be provided with very hygienic living conditions. Kept in too small cages the urge to dominate amongst the high-ranking animals assumes an aggressive character and the lowest-ranking members of the group are subjected to permanent stress, hardly daring to touch the food rations and generally deteriorating physically and psychically so that the group gradually disintegrates from the lowest level. These monkeys are relatively common in zoos. The regionally differing forms are frequently classified into four species with 15 subspecies.

Squirrel Monkey *(Saimiri sciureus)*. The most common and best known species. The head is ocher colored. It occurs in South America to the north of the Amazon; to the west in the region south of the Juruá and Ucayali rivers, and in the east in the Xingú region.

Red-backed Squirrel Monkey *(Saimiri oerstedii)*. The body fur is more brightly colored and tinged with reddish brown, while the head is dark gray. Occurs from Panama to Nicaragua—the most northerly species in Central America.

Madeira River Squirrel Monkey *(Saimiri madeirae)*. Slight deviations from the previously mentioned color scheme. Its range lies to the south of the Amazon, to the east and west of the Madeira river.

Black-headed Squirrel Monkey *(Saimiri boliviensis)*. Also displays slight variations in the color and markings. This species occurs in southern Bolivia to Paraguay and in the northern part of Argentina as far as the Bermejo river.

Capuchins (genus *Cebus*)

Of all the New World monkeys the capuchins display the closest resemblance to the Old World monkeys. As early as 1578 the French explorer J. de Lery spoke of a little black "guenon" after his return from Brazil. Linnaeus, too, grouped this genus together with the guenons and macaques.

Capuchins are medium-sized monkeys with a body measuring from 32 to 56 cm, the tail length varying between 38 and 56 cm; they weigh from 1.1 to 3.5 kilos, in some cases more. The tail, which can be rolled together at the tip, is neither naked on the under-surface nor suitable for gripping in the way the spider monkeys do; it is used exclusively as a means of support. Like the squirrel monkeys, the capuchins' diet is partly vegetarian and partly carnivorous; besides insects they also eat small lizards, amphibians, and small birds. Gestation lasts about 180 days and one young is born which is first of all carried ventrally by its mother and soon after on her back and often on that of the male, too. It moves over to the mother to be suckled. The capuchin becomes reproductive when it is about five years old. A maximum age of almost 47 years has been registered, probably the highest among the monkeys (F. Jantschke 1978).

Capuchins inhabit large parts of South America—the hot damp forests of the Amazon Basin, wooded upland areas to an altitude of 1,600 meters, as well as the tree fringe of savannah or fertile plantation country. In groups numbering from 8 to 30 animals these lively monkeys tour their territory at daybreak in search of food. Quite often the group is scattered over an area of 300 or 400 meters but they remain in constant acoustic con-

tact. When traveling they always keep to a certain order: the troop is headed by young animals, followed by the adult males and females, with the mothers and their young bringing up the rear. The capuchins, too, urinate on their hands and then rub them on their fur and feet. Probably this "scent" is used to mark their most frequented routes. A high-pitched whistling sound denotes alarm, and when agitated or frightened they utter loud squeals. The highest-ranking animal dominates the group; the rest of the hierarchy appears instable. When a member of the group is in danger the dominant males generally challenge the attacker with a threatening stare and bared teeth as well as cries. Greetings and the desire for contact are expressed by the smacking of lips; the capuchins' facial expressions are altogether very highly developed. A state of agitation can also be recognized by the male's erected penis. Much time and care are spent on grooming their own fur and that of other members of the group—another sign of social bonds. Like the squirrel monkeys they display obvious pleasure in rubbing their backs and necks with those parts of plants containing volatile oils—orange peel, for instance. They also use onions for this purpose, causing tears to run down their cheeks. Could this possibly be a help in warding off insects?

The capuchins are remarkably intelligent representatives of the New World monkeys. "Mungo," a capuchin at Dresden Zoo, drew with chalk, adorning the walls of his cage with triangles, spirals and close zig-zag lines. He mixed various ingredients in a bowl with a spoon and even managed—though not always quite successfully—to break an egg into this "cake mixture." He used to hose down his cage too. One of his most remarkable feats was his escape from the outer cage. He took a split-off branch from his climbing tree and used it as a crowbar, systematically working it backward and forward to loosen the binding wire that attached the meshwork to the iron frame. Having completed the first stage of the job, he then prised the meshwork out of its frame, escaping through the resulting long slit. It was a convincing example of a certain degree of foresight and reasoning in the use of tools.

Capuchins are better at experimenting and "inventing" on their own than at imitating. They extract insects from their hiding-places with the aid of stems or rolled-up leaves and they can crack nuts with stones or other hard objects. Because of their notable mental performance, these monkeys were already used in research laboratories many years ago and their engaging characteristics made them popular domestic pets with the additional advantage that they required no extreme care. Some remarkable successes have also been achieved with them as performing animals. The capuchins are amongst the most frequently exhibited New World monkeys in zoos and are quite often bred in captivity. In 1977 the world's zoos registered a total of 79 births.

Nowadays the multitude of forms embodied by the capuchins is classified into four species with many subspecies. Because of the many varieties and transitional forms an exact classification is very difficult.

White-throated Capuchin *(Cebus capucinus)*. Relatively small, slender, dark brown in colour with white shoulders and face; short-haired. Range: Central America from Guatemala to northern Colombia.

White-fronted Capuchin *(Cebus albifrons)*. Slender body, relatively small; medium brown fur, dark at the top of the head, white forehead and face. Range: Colombia, Venezuela, on Trinidad, in northwestern Brazil and northern Bolivia.

Weeper Capuchin *(Cebus nigrivittatus)*. Somewhat larger body and elongated limbs; medium brown shaggier fur; short-haired head, whitish forehead and sides of head, pink and naked face. Found in Guayana, Surinam to the mouth of the Orinoco, and in Brazil north of the Amazon from its delta to its confluence with the Río Negro.

Apella or **Fawn Monkey** *(Cebus apella)*. Body more powerful and compact, shorter limbs, medium to dark brown longish fur. Top of head dark to black with various "hair styles." Longer tufted hair forms two crests over the ears. Occurs in disjunct areas extending from western Venezuela over eastern Colombia, western Brazil, Bolivia to the southernmost part of Brazil; they are also found south and north of the mouth of the Amazon. More widely distributed than any other capuchin species.

Howler Monkeys (subfamily Alouattinae)

As the name indicates, these are exceptionally noisy animals. The large hyoid bone distends in the form of a hollow shell, greatly amplifying the sound emitted. The shape of the lower jaw which has resulted from the enlarged hyoid bone gives the head a strange appearance in profile. Weighing between 7 and 9 kilos the howler monkeys are the heaviest New World primates; their bodies measure from 56 to 60 cm and their tails are from 60 to 70 cm long. Like the spider monkeys, the muscular tail is prehensile, a so-called "fifth limb," its tip being furnished with a naked surface for about 20 cm to give a firm grip. The powerful thickset body is covered with long silky fur which varies in color according to species, subspecies, age and sex. All species have naked faces. In their movements the howler monkeys are deliberate and slow and do not risk big leaps from branch to branch. Like all semi-brachiators they can also move about by swinging arm over arm from bough to bough and rarely come down to the ground. Their diet is mainly composed of

Howler monkey *(Alouatta palliata)* feeding (after Dorst).

Howler monkeys (*Alouatta* spec.)

50 different kinds of leaves, young shoots, blossoms, and fruit. They have hardly ever been known to eat food which is of animal origin.

Gestation lasts four and a half or five months; the newborn weighs about 600 grams and is silvery gray in coloring. For the first four weeks it is carried ventrally and afterwards on the back of its mother. The young become independent when they are about three years old.

The first exact behavioral studies of howler monkeys in their natural environment were undertaken by C.R. Carpenter on the Barro Colorado Island (Lake Gatun) in Panama in the nineteen-thirties. Similar studies have also been conducted more recently, supplying information about the well developed and interesting social behavior of these monkeys. They move about in specific, clearly defined virgin forests in groups of from 14 to 20 animals (approximately three males, seven females, three children and four young animals). The chief features of a group's territory are the feeding and sleeping places which are connected by regular arboreal "routes." The monkeys behave amicably to each other. Attackers are challenged by the males. In their sexual behavior none of the males appears to play a dominant role. Sexually receptive females solicit the attention of a chosen mate by rhythmic motions of the tongue between the oval-shaped lips. If the male is willing, he responds in a similar fashion. Sexually potent females mate with several males. Members of a group communicate with an extensive range of

calls in which deeper notes predominate. Their loud choruses at dawn can be heard for miles. The male has a more powerful voice than the female. The calls help troops that are far apart to keep in touch or, alternatively, convey a warning to those in the proximity to keep at a distance in order to avoid territorial conflicts. If, however, two groups do confront each other, regular "howling contests" take place; the noisiest and most persistent group is the winner. Intruders on the ground are driven off by a bombardment of branches.

In zoological gardens howler monkeys are today still regarded as difficult charges with a relatively brief life expectancy. This is possibly due to their complicated specialized diet.

The one genus contains five or six species with many subspecies and varieties.

Howler Monkeys (genus *Alouatta*)

Guatemala Howler Monkey (*Alouatta villosa*). Black, soft and silky fur, naked face. Found in a very restricted habitat in northern Guatemala.

Mantled Howler Monkey (*Alouatta palliata*). Blackish brown fur, often lighter and longer on the sides of the body. Range: southern Mexico round the Bay of Campeche; Central America from Belize to the Pacific coastlands of Colombia.

Red Howler Monkey *(Alouatta seniculus).* Larger than the other species, with mainly red or reddish brown fur. The subspecies vary a good deal in coloring and markings. Range: eastern Colombia, western and eastern Venezuela over Surinam to the mouth of the Amazon and in northern Brazil.

Brown Howler Monkey *(Alouatta fusca).* Brown to reddish brown fur, face surrounded by a beard. Range: southeastern Brazil and a small zone in western Bolivia to southeastern Peru (?).

Rufous-handed Howler Monkey *(Alouatta belzebul).* Black to dark brown fur; light reddish brown feet and tail; there are also yellowish to reddish variations. Range: south of the Amazon from its mouth to the lower course of the Juruá and in the south to about the 10th degree of latitude.

Black Howler Monkey *(Alouatta caraya).* Larger body. The coat of the male is black, that of the female and young animals yellowish to olive-brown; thickly bearded. Range: northern Argentina, Paraguay, eastern Bolivia and southwestern Brazil.

Spider Monkeys (subfamily Atelinae)

This subfamily is one of the most typical groups of South American primates and, along with the howler monkeys, the largest in size. The body is generally slender, the limbs long and thin. The prehensile tail is long, powerful, muscular, mobile and tactile. As a grasping organ it functions as well as a hand. The long arms, legs and tail are used for swinging from bough to bough—a method of locomotion known as brachiation. The spider monkeys are only excelled in speed, acrobatic abilities and elegance by the Asian gibbons. When sitting or lying down the tail serves as a safety mechanism. Hanging by their tails they have both hands free for gathering food. While at play in Dresden Zoo three young spider monkeys were seen to clutch the body of another one that was hanging by its tail, so that four of them were swinging on one tail. The hands are narrow and elongated, the thumbs have almost completely disappeared.

Woolly Monkeys (genus *Lagothrix*)

The woolly monkey, named after its thick wool-like and soft fur, makes a somewhat thickset and "athletic" impression. Its head and body measurements range from 55 to 65 cm; the muscular tail is 60 to 70 cm long; a specimen can weigh up to 7 kilos. The thumbs are not noticeably reduced in size. Woolly monkeys normally move deliberately, steadying themselves with their prehensile tails, in groups of from 12 to 30, rarely more than 50

animals, along the branches of the rain forest. But by impending danger they become very nimble, reaching out and swinging adeptly from branch to branch and even risking a downward leap into the dense foliage below. But first of all they attempt to hide.

The diet comprises fruit, leaves, sometimes insects and other small animals too. They use the tips of their tails to angle fruit hanging on thin stalks within picking range. Woolly monkeys have a very well-nourished appearance—in captivity they are omnivorous—and the Brazilians call them "barrigudo"—potbelly. Only one baby is born at a time and it is suckled by the mother for about a year. The sense of smell plays an important part in mutual communication; they rub their chest glands on twigs and smell each other's chests which the male usually salivates. Because of their amicable disposition they often share the housing of capuchins, spider and howler monkeys in zoos. Tame woolly monkeys are frequently kept by the local people, but are also often eaten by them. In captivity they are fastidious charges. But with careful handling even private owners have managed to breed them.

Like other New World monkeys they vary in appearance locally so that a systematic classification into subspecies is extremely complicated. Nowadays only two species are classified:

Woolly Monkey *(Lagothrix lagothricha).* The coat is mainly pale brown, light gray to blackish in colour, with the head mostly darker and round in shape. **Range:** Central and upper Amazon Basin, in a northerly direction as far as Colombia, in the south extending to Bolivia and in the west to the foothills of the Andes. There are four subspecies: *Lagothrix lagothricha cana; Lagothrix lagothricha lagothricha; Lagothrix lagothricha poeppigii;* and *Lagothrix lagothricha lugens.*

Peruvian Mountain Woolly Monkey *(Lagothrix flavicauda).* The body is deep mahogany colored, darker in front; the under-surface of the tail is yellowish from the middle to the tip with a distinct division from the upper surface; there is a distinct brownish yellow triangle from the upper lip to the root of the nose; the scrotum is marked by a thick yellow tuft of hair. The range is restricted to a limited area in the eastern forelands of the Peruvian Andes.

Woolly Spider Monkeys (genus *Brachyteles*)

Despite their close relationship with the spider monkeys, they are classified as a full genus with one species.

Woolly Spider Monkey *(Brachyteles arachnoides).* This animal has conspicuously elongated limbs. Its body length ranges from 48 to 64 cm, its tail may measure up to 89 cm. In exceptional cases it may weigh 10 or 11 kilos. The thick, woolly, yellowish gray-brown coat is overlain and interspersed with coarser hairs. The thumb is much reduced in size or non-existent; the other fingers have long nails. The naked grayish pink to red face stands out against the drab coat. There is a fringe of black hair over the eyes. In profile the face appears angular. The long prehensile tail is very powerful. According to observations made by H. Wendt these monkeys are slower and more deliberate in their movements, forming a marked contrast to the very lively spider monkeys. In the wild they are very shy and even Indios on the hunt rarely catch a glimpse of them.

Their range in southern Brazil between the upper course of the Paraná and the coast was originally covered by dense virgin forest. Today largely opened up for cultivation, this area, with the exception of the Tupi forests, barely affords habitats for these monkeys so that their existence is severely threatened.

Very little is known about their habits in the wild. In zoos, even in their homeland, they are considered to be great rarities. Classified by the IUCN as a species threatened with extinction.

Spider Monkeys (genus *Ateles*)

The spider monkeys represent the most typical long-armed "tail acrobats." I. T. Sanderson wrote that movement through the tree-tops is typical; they remind one of an acrobat on a flying trapeze; one hand grips a branch and, hanging by the arm belonging to it, the body swings forward through space, the hand releases its grip, the body flies away, the other hand is extended, grips a new support and then, suspended on its arm, the body swoops off again, and so on. In this way spider monkeys can hurl themselves up to 10 meters across space, crossing the tree-tops as rapidly as a man can run across an uninterrupted stretch of flat ground. The tail may exceed the maximum body length of 79 cm by an average 25 cm. The naked under-surface of the end of the tail displays characteristic papillary ridges which vary individually. In determining the sex it is important to note that the female has a long peniform clitoris whereas the masculine penis is very small and barely discernible. Another deviation is the location of the nipples close to the armpits. The glands near the breast-bone play a stimulating role in courtship. After 140 days' gestation, one, more rarely two, offspring are born. Newly born are found all the year round, although there is probably a season when the most births take place. Because of their missing thumbs and the resulting loss of manipulative power, grooming is not a pronounced activity amongst the spider monkeys.

Their diet is predominantly composed of fruit, although, according to H.O. Wagner, they also rob birds' nests. They find nectar in the dendrophytus blossoms that they tear off and open. They bite into the *Tocoyena pittieri* fruit, but leave it on the stalk when the pulp is still green and bitter. This opening of the fruit accelerates its ripening process. When the monkeys pass the same way a day or two later they can feast on the syrupy sweet pulp (Hladik 1969). Spider monkeys have amazingly good memories. Even after years have gone by they still recognize people with whom they once had good contact. The generally amicable spider monkeys live in groups of from 10 to 40 animals. They defend their territory against intruders, but rarely present an opportunity for attack. They prefer a discreet retreat whereby the group splits up into smaller troops. Generally speaking, the various groups respect each other's territory so that ritualized battles rarely occur. Within the group itself, about a third of which is made up of males, the females play a more active and leading role in comparison to other species. Fairly often subgroups are formed, mostly consisting of a sexually receptive female and one or more males. Superfluous males do not go off on their own but form masculine groups. Nowadays spider monkeys are being kept with increasing success in zoos and many births in captivity have been registered. Sometimes the foetus is very large and dangerous complications occur when it is born. The spider monkeys' coloring and markings are very variegated. Even within one group there are often distinct variations although mostly one of these predominates. There are four species with several subspecies:

Central American or **Geoffroy's Spider Monkey** *(Ateles geoffroyi)*. Predominantly pale to medium brown coloring on the body, darker fur on the limbs. Occurs in Central America from southern Mexico to Panama. One of the subspecies is *Ateles geoffroyi panamensis*.

Brown-headed Spider Monkey *(Ateles fusciceps)*. Body predominantly black. Occurs in very restricted areas in the Pacific coastal region of Colombia up to the Gulf of Darién.

Long-haired Spider Monkey *(Ateles belzebuth)*. A black body and dark face with flesh-colored patches round the mouth, nose and eyes; white to yellow stripe on the forehead, white cheek stripes *(Ateles belzebuth marginatus)*. Other forms have pale gray to yellowish undersides, extremities and under-surface of the tail *(Ateles belzebuth belzebuth)*. Found in Central and eastern Colombia, in the south bordering on northern Peru and Ecuador; in the north to the southwest of the Gulf of Venezuela and Lake Maracaibo as well as to the south of the mouth of the Amazon as far as the lower course of the Tapajós.

Black Spider Monkey *(Ateles paniscus)*. Black body, flesh-colored face *(Ateles paniscus paniscus)*; with black face *(Ateles paniscus chamek)*. Occurs in two disjunct areas: Guayana, Surinam, French Guiana and southward to the lower course of the Amazon, as well as in western Brazil to the south of the Amazon as far as northeastern Bolivia.

Family Callimiconidae

The discovery of this family has a somewhat unusual history. The Swiss zoologist E. A. Goeldi, originally director of the Brazilian State Museum of Natural History and Ethnology in Pará (now Belém), received a small unidentified monkey of uncertain origin. The British Museum in London classified it as a new species of titi monkey on the basis of the fur of the now dead animal—the skull was missing—naming it after Goeldi. Later on, when two further complete bodies of this strange monkey were examined, the Brazilian zoologist M. Ribeiro identified it as an intermediate form linking the marmosets and tamarins (anatomy, claws, fur) and capuchins (structure of the skull and teeth). It then received the new genus name of *Callimico* (pretty little monkey). The zoologists have so far failed to agree upon its definitive classification. Napier and Napier (1967) placed this "outsider" in a special subfamily of the marmosets and tamarins. Today this "double-track" monkey is usually allotted the status of a full family. Before detailed knowledge about it was acquired, the marmosets and tamarins were regarded as the most ancestral "pre-monkeys," situated far down the ladder of primate evolution. In the meantime it has been proven beyond a doubt that the marmosets and tamarins are a dwarf branch of the flat-nosed monkeys. The lengthy controversy about the little black monkey thus exerted considerable influence on the ultimate classification of the marmosets and tamarins.

Genus *Callimico*

Goeldi's Monkey *(Callimico goeldii)*. Body length from about 24 to 27 cm; length of tail ca. 26–32 cm. Weight: 450 to 500 grams. It has a rounded skull although the jaw region is somewhat jutting. The large, round, naked ears are almost hidden by the upstanding hair on the head and a longish mane in the nape of the neck. The soft and silky fur is jet black with a brown to golden shimmer, partly caused by lighter-tipped hair. Behind its base the tail is faintly marked with several indistinct pale sandy rings. The long fur on the rump forms a kind of "skirt" when spread out over the base of the tail. The hind limbs are longer than the fore limbs. The toes and fingers have narrow, claw-like, elongated nails, only the big toe having a flat nail. The short thumb is opposable to the other fingers. There are 36 teeth and the dental formula is $\frac{2 \cdot 1 \cdot 3 \cdot 3}{2 \cdot 1 \cdot 3 \cdot 3}$. The animal loses its first teeth at the early age of nine months. The offspring are born one at a time after gestation lasting from four and a half to five months and are suckled for about 60 or 70 days. The diet consists chiefly of fruit but also includes insects, birds' eggs, small vertebrates.

These rare monkeys occur in four disjunct areas in the rain forests of Peru along the upper course of the Amazon between the Marañon and Ucayali. Little or almost nothing is known about their habits in the wild. Our knowledge has so far been acquired through studying their behavior in captivity. Lively by disposition, they can become very tame. They have an extremely well developed leaping ability. With a few exceptions specimens in zoological gardens have so far not survived very long.

Marmosets and Tamarins (family Callithricidae)

This large family of neotropical monkeys does not represent an ancestral form but is a result of pronounced specialization, in the course of which very divergent and unusual varieties came into being. Without exception, they are all small animals that hardly ever leave the trees. Natural barriers, such as rivers that with the passage of time changed their course more than once, mountain ranges, savannahs, grasslands and barren country have in many cases repeatedly isolated the groups, thus producing a multitude of forms.

With a body length of from 16 to 31 cm, the tail varies between 18 and 42 cm and the weight from 80 to 570 grams. The silky and often conspicuously marked or colored fur is adorned by manes on the head, shoulders or backs of some of the species. The complete adaptation of the marmosets and tamarins to an arboreal life is indicated, for instance, by their claw-like nails that are not an ancestral feature since the embryo begins to develop monkey nails that later change into claws. The foot is developed as a grasping organ and the big toe has a broad nail. The short jaws leave no room for the last molars; there are only 32 teeth and the dental formula is $\frac{2 \cdot 1 \cdot 3 \cdot 2}{2 \cdot 1 \cdot 3 \cdot 2}$. The second teeth appear when the animal is only seven to nine months old.

Marmosets and tamarins utter very high-pitched calls with considerable ultrasonic frequency and a variety of bird-like sounds such as twittering, cheeping and screeching. Gestation lasts from 140 to 160 days and twins are usually born whose eyes open on the first day. The male takes a very active part in rearing the young which are suckled up to three months. Compared with other species, marmosets and tamarins have a brief life expectancy of from 12 to 15 years, rarely living longer. They feed on both plants and animals.

These monkeys are usually found in the rain forests where they generally keep to the upper regions, "the woods over the woods" as Alexander von Humboldt aptly described the dense vegetation cover consisting of epiphytes, ferns, lianas and other creepers. As a result little is known about their habits when resident in the wild.

All marmosets and tamarins have a social way of life and generally form hierarchical groups of ten or more individuals. The order of precedence is decided by battles in the course of which severe injuries may be sustained. On the other hand, the social groups of various species coexist amicably. These otherwise very lively and mobile animals often huddle close together in small groups on a branch when at rest. The graceful little monkeys are very adept hunters of small animals, killing their prey swiftly by a paralyzing bite in the head.

During recent years behaviorists have started to pay more attention to these New World monkeys. Their studies are expected to produce results of considerable comparative value in view of the substantial differences that exist between these and other monkeys.

Marmosets and tamarins were already tamed and kept as domestic pets by the indigenous inhabitants of South America. With the discovery of the New World these dainty little monkeys were soon brought to Europe. As early as 1551—about sixty years after they had been discovered by Europeans—K. Gesner (Swiss physician and naturalist) portrayed the lion marmoset and the common marmoset in his *Historia animalium*. Since then many of these monkeys have lived in captivity. Fashionable ladies used to carry them in their sleeves. In Baroque days it was customary amongst the French aristocracy to present a wife or mistress with one of these dainty dwarf monkeys. La Condamine, a French explorer of South America and discoverer of the caoutchouc tree, brought back the first wigged marmoset and presented it to the French naturalist Comte de Buffon. Provided they were well looked after these monkeys thrived and bred satisfactorily both in zoos and in private ownership. Because of their popularity and suitability as laboratory animals they were still being captured and exported in large numbers until fairly recently so that several species have now been registered by the IUCN as rare or endangered.

The multitude of forms, specific localized variations and in some cases the different appearance of the sexes make it extremely complicated to define the species and subspecies. Moreover, the geographical distribution of the majority of forms is so far only partly known. Consequently opinions differ about their classification. Taking the teeth as a point of departure the following distinction can be made: **Marmosets:** lower jaw canine teeth and incisors of almost equal length. **Tamarins:** lower jaw canine teeth much longer than the incisors.

Marmosets (genus *Callithrix*)

The body length of these little monkeys ranges between 20 and 25 cm; the tail is from 29 to 35 cm long. There are considerable variations in the color and markings of the fur even within the species and subspecies. Most of the species are furnished with conspicuous fan or pen-shaped bushy ear tufts. Because all the fingers are on one plane the marmosets' grip appears clumsy, but they clamber adeptly in the branches. In order to lick up juicy sap they gnaw holes with their incisors in the bark of trees.

They live in pairs, together with their offspring, forming small groups. Scent from the chest glands plays a major communicative role in courtship. The displaying of genitals is a visual indication of sex from which the dominance status can be deduced. Stereotype calls are frequently heard a long way off. The males are extremely active in looking after the young. Soon after birth the offspring are carried around by the father and only move briefly to the mother for suckling. They gradually abandon their "pick-a-back" rides when they have learnt to move on their own.

Common Marmoset (*Callithrix jacchus*). The thick, soft, long-haired fur of this marmoset has brown, yellowish white and orange tinges. Lying sleekly against the body it appears cross-striped, and ringed on the tail. There is a white patch on the forehead.

These marmosets live in groups which are usually dominated by a socially high-ranking male and female. The individual members of the group demonstrate their social status with specific display behavior such as threatening grimaces, bristling fur, waggling ears and bowing and scraping. The scent glands in the genital region have an important function, furnishing an olfactory signal about the social status of high-ranking animals. As with most of the monkeys, mutual grooming forms a prominent activity in their social life.

Their area of distribution is eastern Brazil where they are still frequently found between the 3rd and 23rd degrees of southern latitude in the rain and dry forests, in tree savannahs as well as in the Tupi forests. They have also been brought to other regions and even occasionally occur in cultivated areas such as parks, suburban gardens or plantations. Breeding successes in captivity have substantially increased over recent decades.

Buff-headed Marmoset (*Callithrix flaviceps*). The white region round the mouth stands out against the yellowish brown head. The yellowish white ear tufts are fan-shaped and somewhat shorter than those of the other tufted species. The habitat more or less coincides with that of the common marmoset. But this species is very rare; some zoologists class it as a subspecies.

Black-penciled Marmoset (*Callithrix penicillata*). As the name indicates, dark brown to black drooping pencil-shaped ear tufts adorn the head. Head and neck are dark brown, the body predominantly gray and there is a white triangular patch on the forehead. This species occurs in a zone extending from the plains in southeastern Brazil to approximately 28 degrees southern latitude. Some zoologists classify the following four forms of this group just as subspecies.

White-eared Marmoset (*Callithrix aurita*). Similar to the black-penciled marmoset in appearance. The abdomen is, however,

Pygmy marmosets *(Callithrix pygmaeus)*

ocher, hands and feet yellowish, while the ear tufts vary between white and gray. Its habitat approximately corresponds to that of the black-penciled marmoset. Classified by the IUCN as a rare species, only 12 specimens in four zoological gardens were registered in 1978.

White-fronted Marmoset *(Callithrix leucocephala (C. geoffroyi))*. Distinctive features: upper side of the body ocher with blackish streaks and yellowish white spots; underside black to brown, throat and chest white, black mane in the nape, face covered with short white hairs. Found in southeastern Brazil, especially in the wooded areas in eastern Minas Gerais and in Espírito Santo.

White-shouldered Marmoset *(Callithrix humeralifer)*. The back and sides of the body are blackish brown with a partially gray tinge; abdomen, arms, shoulders and throat are white, the ear tufts black. The animals are also found in eastern Brazil, especially in the forests of eastern Bahía in the vicinity of El Salvador.

White-necked Marmoset *(Callithrix albicollis)*. General resemblance to the white-fronted marmoset but with yellowish white ear tufts and a yellow to white neck. According to studies made at Göttingen University only the highest ranking female

and male in larger groups of this species are fertile—at least when kept in cages. The lower-ranking animals are stressed to an extent that produces "psychological sterilization."

Santarem Marmoset *(Callithrix santaremensis)*. Less well-known than the previously mentioned species. The relatively long legs indicate leaping ability. Its range extends to the south of the middle reaches of the Amazon approximately between the rivers Purús and Javari, covering a stretch of about 400 kilometers.

Yellow-legged Marmoset *(Callithrix chrysoleucos)*. White fur with a silky gloss, back and crest varying between yellow, light brown and ocher; hands, feet and tail are yellow to golden ocher. It is sympatric with the previous species.

Silvery Marmoset *(Callithrix argentata)*. Differs in habits and appearance from the other marmosets and is very peaceable by nature. As indicated by its stronger canine teeth it eats more food of animal origin. Some Indios and mestizes believe that its bite is venomous. On the other hand, when tamed, it is prized as an affectionate domestic pet. In captivity breeding successes have often been registered. Weight at birth is between 26 and 28 grams. There are several subspecies differing substantially in appearance from the other species:

Black-tailed Silvery Marmoset *(Callithrix argentata argentata)*. The best-known form has a long-haired, thick and silky silvery white coat, while the short-haired tail is black. The reddish yellow face, head and the large ears are naked. It lives in the forests on the southern banks of the lower reaches of the Amazon. The two other subspecies, *Callithrix argentata melanura*, whose body coloring is more yellow and brown, and *Callithrix argentata emiliae*, which has a silvery gray to light brownish gray coat, occur in southern Mato Grosso and southern Pará.

Pygmy Marmosets (subgenus *Cebuella*)

Pygmy Marmoset *(Callithrix pygmaea)*. First discovered by J. Spix near the upper course of the Amazon in 1823, the pygmy marmosets were originally thought to be the offspring of common marmosets. Only one species exists. With a body length of only 16 cm, a tail measuring 18 cm and weighing from 80 to 95 grams, these are the smallest monkeys (the smallest primate of all is the mouse lemur—see page 31). The fur is greenish yellow with pale dark brown diagonal stripes; the ears are hidden in the long mane and have no tufts. Troops of these animals move swiftly and nimbly along the branches of the rain forest. By impending danger they hide behind an upright tree trunk, keeping this between themselves and the potential enemy. Their repertory of calls resembles the twittering of birds and the chirping of crickets. Some sounds are beyond our range of hearing.

Their diet consists of fruit, insects, and small vertebrates. They occur in a region stretching from the north of the upper course of the Amazon, along the forested banks of the Napo in northeastern Peru to southeastern Colombia. After gestation lasting four months identical twins are usually born, measuring about 3 cm at birth and ca. 4.5 cm after a week. They are carried around by the father and only transfer to the mother for suckling. The offspring start to move independently when their joint weight is approximately equivalent to that of their father. They are suckled for about a month. After the Second World War larger numbers of these pygmy marmosets were brought to zoos in the USA and Europe and also became popular pets for private owners. Little is known about their wild-life habits. Breeding successes have been achieved in recent years.

Tamarins

In contrast to the marmosets the tamarins constitute a very diversified and extensive group. Its representatives are somewhat larger, their longer limbs make them better at jumping and their dentition differs from that of the marmosets. There are three genera:

Maned Tamarins (genus *Leontideus*)

Their exact classification is still a matter of dispute since some of the species do not possess the typical tamarin features. The coloring is very conspicuous. The yellow to reddish gold long-haired fur has a soft and silky sheen and lies loosely on the body. Head and neck are adorned with a mane which has given this genus its name. The largest amongst the Callithricidae, these animals have a body length of from 28 to 35 cm, the tail being about 40 cm long. The sparsely haired face is of a pale purplish mauve color. The pointed canine teeth add to the impression of a mini-lion's head. Amazingly elongated extremities, whose fingers are almost as long as the forearm, make it possible for them to jump great distances. A skin fold on the third fingers gives the nimble and adept animals a sure hold as they run along the twigs. Living in small groups they prefer the upper regions of the dense rain forest. Their calls and diet correspond more or less with those of the other marmosets and tamarins.

They give birth to twins which are carried in the mother's fur to begin with but which transfer to the father after five to nine days. At feeding-time, about every two hours, the male leans sideways toward the female and gives her, with the partial aid of his hands, the offspring for about 15 minutes. Ditmar states that the father crushes soft bananas with his hands, thus accustoming the young to solids. After about five months, when they have already made their first efforts to clamber, they begin to fend for themselves, seeking refuge in the parents' fur only by impending danger.

In captivity these monkeys exhibit a definite need for contact with their keepers, whom they often regard as substitute partners. Sanderson underscores their good memories and intelligent behavior, particularly their astounding ability to distinguish between voices, sounds, etc.

Probably these monkeys once inhabited a larger territory encompassing the area from Bahía to São Paulo as far as the Río Paraná. Extensive forest clearance, cultivation and other economic encroachments have substantially diminished their habitat. Since they do not settle in cultivated areas they now only occur in remote small areas in which the herds are alarmingly threatened. All the species that are endangered have been registered as such by the IUCN and placed under strict protection. The international studbook is kept by the Washington National Zoological Park. Formerly occasionally seen in zoos, they are today numbered amongst the very rare exhibits. The three species are also sometimes classified just as subspecies.

Golden Lion Marmoset *(Leontideus rosalia)*. Coat almost entirely golden orange. Occurs in the forests of the coastal uplands to the southwest of Río de Janeiro up to an altitude of 1,000 meters.

Golden-headed Tamarin *(Leontideus chrysomelas)*. The golden orange fur is only found on the head, neck, upper arms, loins and upper surface of the tail, the other parts of the body being black, sometimes with a reddish tinge. Only occurs in a very small coastal region of southeastern Bahía. Very rare.

Golden-rumped Tamarin *(Leontideus chrysopygus)*. A very long mane, upper part of the head golden orange; black trunk, hands, and feet, only the lower part of the back, flanks and upper part of the tail being a rusty red. Found locally in a few small areas in southeastern São Paulo. Very rare.

True Tamarins (genus *Saguinus*)

With a body length varying between only 25 and 31 cm, the tail is from 30 to 42 cm long and the weight ranges between 300 and 560 grams.

Negro Tamarins (subgenus *Saguinus*)

They belong to the Callithricidae that were discovered at an early date and are mentioned in the writings of Gesner, Linnaeus, Buffon and other naturalists. They seek newly grown tree cover (secondary forests) and arboreal plantations to live in and are usually found in the tree-tops. Field studies have confirmed that they also coexist amicably with capuchins. By impending danger they retreat with alarm calls and scolding sounds to the most inaccessible tree-tops or remain motionless hidden in the canopy of twigs. Like other species, they are often captured by the local people and kept as very affectionate and docile pets. Those that escape from captivity are then fairly often encountered in the gardens of settlements and town parks. In recent years there have been more reports of breeding successes by various zoos.

Negro Tamarin *(Saguinus tamarin)*. Has a jet black coat with a more or less clearly marked gray to ocher-colored saddle. The naked face is also gray. Its range in northeastern Brazil extends from the lower reaches of the Tocantins to Surinam. These tamarins are still found occasionally in the surburbs of the port of Belém (formerly Pará) and are therefore also known as "Pará monkeys." The herds are, nevertheless, much depleted.

Red-handed Tamarin *(Saguinus midas)*. Similar coloring to the previous species, the hands and feet, which are red, orange, ocher-yellow or chestnut brown according to subspecies, forming a distinct contrast to the black coat. It occurs in an area north of the lower reaches of the Amazon as far as Guayana, particularly in the upland forests.

Moustached Tamarins (subgenus *Tamarinus*)

An unmistakable feature of these somewhat smaller and daintier animals is the white beard-like hair above the upper lip and round the mouth. Twelve species and twice that number of geographical subspecies have been established. An exact classification is often complicated by the extreme variability within a single species. The reason for this may be based on sex, age, origin, territory, specific types of coloring, etc. In their disposition and repertory of calls they resemble the penciled marmosets. They inhabit the vast, and still incompletely explored virgin forests of the upper Amazon Basin. As a result of this inaccessibility and lack of development projects at that time, many species were only discovered by zoologists at the beginning of the 20th century. The Indios often tame these animals because they are adept at picking parasites out of the hair of their owners. As far as ascertainable, the numbers of most species are still stable, so with a few exceptions they are relatively common exhibits in zoological gardens where the majority of species have bred several times.

Black and Red Tamarin *(Saguinus nigricollis)*. A thin white moustache that looks as if it had been dipped in milk; dorsal side black to dark brown, lighter towards the back, mottled reddish brown hindquarters. Like most species, it is difficult to define the limits of their range. Fairly frequently exhibited in zoos.

Brown-headed Tamarin *(Saguinus fuscicollis)*. Hard to classify as a result of many variations in the color scheme. The head, neck, nape and front of the arms are mostly dark brown to black, white "eyebrows," mottled black and reddish yellow or rusty red back, pale brown hindquarters and legs, sometimes darker fur on the middle of the body. Markings often faded.

The **White-lipped Tamarin** *(Saguinus weddeli)*, **Golden-mantled Tamarin** *(Saguinus tripartitus)* and **Red-mantled Tamarin** *(Saguinus illigeri)* are other similar species; in recent years American and British zoos have reported several breeding successes with the latter species.

White Tamarin *(Saguinus melanoleucus)*. The unusual coloring—white with dark gray to black face—does not conceal the very close relationship with the brown-headed tamarin. A specimen kept in Cologne between 1965 and 1967 was probably only the second of its kind to reach Europe alive. Very rare.

Moustached Tamarin *(Saguinus mystax)*. In its snow-white face there is a distinctive tuft of white hair round the mouth that, according to Bates, looks like a flake of pure white cotton; the dorsal side is a mottled black, yellow, brown and gray, with substantial variations; underside black to brown; the mane, throat, limbs and tail are glossy black. It occurs from the Andes to the interior of western Brazil.

Lönnberg's Tamarin (*Saguinus pluto*) and the **Rio Napo Tamarin** (*Saguinus graellsi)* are two very similar species.

Red-capped Tamarin (*Saguinus pileatus*). Distinctive red patches of fur, cinnamon-colored crest.

Red-bellied Tamarin (*Saguinus labiatus*). Blackish brown back and head, belly fur a glossy deep to orange red. The face is adorned with a short white moustache. Has been bred several times in zoological gardens.

Imperial Tamarin (*Saguinus imperator*). First discovered in 1907. Mainly black body, tail reddish brown. The body length rarely exceeds 25 cm. In captivity it is docile and affectionate. There are two subspecies that occur in western Brazil in the region between the rivers Purús, Jurúa, Acre, and possibly as far as the eastern borders of Peru. Rare. Only exhibited in a few zoos.

Pied Tamarins (subgenus *Marikina*)

Their appearance differs considerably from that of the other species, principally with regard to the naked or bare black or brown-spotted head with large jug-handle ears that have dark brown or bluish spots. Their habits, disposition and readiness to make friends with human beings resemble those of the previously mentioned species. In their active periods they have a remarkable range of facial expressions. If agitated, they can also turn aggressive. When living in the wild they appear to form well-knit groups with an order of precedence. They occur to the north of the Amazon from the Río Negro to the eastern Colombian borderlands. Only one species with three subspecies is known.

Pied Tamarin (*Saguinus bicolor*). Registered by the IUCN as rare.

Saguinus bicolor bicolor: Ventral side white, extending behind the shoulder region, with an abrupt change to the dark brown to yellowish color of the dorsal side. Found in the Río Negro zone.

Saguinus bicolor martinsi: Dark to olive brown fur, lighter on the arms, legs and abdomen; white upper surface of the hands and feet. They occur as far as southern Guayana.

Saguinus bicolor ochraceus: Monochrome coloring, lighter, paler yellowish brown. It occurs mostly west of the Río Negro.

Crested or **Bare-faced Tamarins** (genus *Oedipomidas*)

Generally speaking, somewhat larger than the true tamarins. Some zoologists regard them as close relations of the pied tamarins. Because of their wig-like crests they are also known in the USA as "cotton-heads" or "cotton-tops." With their long legs and very long tails they are extremely good and agile jumpers. Like Goeldi's monkey they can effortlessly jump a distance of more than three meters from branch to branch. This ability is not only important when fleeing but also for catching birds. A more carnivorous diet is indicated by their comparatively strong teeth with large canines capable of killing prey by a single bite in the head. Since the juvenile animals learn this specific behavior from the adults it would not appear to be instinctive.

Some species have a repertory of calls that sound like the song of birds. They are the only Callithricidae to inhabit regions on the far side of the Andes.

They live in larger family groups headed by a powerful and experienced male whose dominant status is demonstrated and defended with specific threatening gestures. The play of facial expressions has a definite meaning and is addressed directly to the partner or enemy. When threatening, the head is lowered, the lips protrude and the mane on the head and nape stands on end. A fixed stare at the opponent denotes an imminent attack. As a preliminary to mating the female rubs her tail with the stimulating scent secreted by her genital glands. She also animates the male by licking his face.

In captivity they require a varied diet of animal origin containing hair, feathers and chitin as bulk material, in addition to the usual vegetarian fare. They were amongst the first South American monkeys to be brought to Europe.

Cotton-head Tamarin (*Oedipomidas oedipus*). Olive-brown on the dorsal side, white chest, abdomen and extremities, reddish brown tail, dark gray face. The family groups number between 8 and 12 individuals and consist of the parents and offspring of various ages, according to H. Wendt. Only the highest-ranking female produces offspring, all the others living under a kind of "sexual stress." When she disappears from the scene her place is immediately taken by the next female who is soon with young. In connection with the sexual cycle—here too at monthly intervals—only the highest-ranking female marks her territory by rubbing the pungent scent of her genital glands on twigs and branches. Gestation is reported to take between ca. 140 and 170 days.

This species occurs in only a very small area of Tierra caliente on the Colombian Caribbean coast. It is thought that they once also inhabited the forests of Ecuador's Pacific coast. Unfortunately the herds are becoming more and more depleted. In zoos they are still relatively common, also in larger groups.

Panama Wigged Tamarin *(Oedipomidas geoffroyi)*. The dorsal side is brown, dark brown to black with stripe-like white patches; the second half of the tail is black, the ventral side white and the face dark gray. Found in the eastern part of Panama to the Colombian Pacific coastal region. This species, too, is unfortunately losing more and more of its habitat as a result of human encroachments. A nature reservation for these tamarins has been established on the Barro Colorado Island in Lake Gatun (Panama). Their habits are similar to those of the cotton-head tamarins. They are uncommon in captivity but several zoo breeding successes have been recorded.

White-footed Tamarin *(Oedipomidas leucopus)*. Although its classification has not been finally settled it is included here because of its great resemblance to the cotton-head tamarin. Its silky fur is light grayish brown on the dorsal side and a rusty red underneath; hands and feet are white. Instead of a mane it only has a white crest on the top of the head. This rare and little-known monkey lives in the river valley forests between the Eastern and Western Cordilleras in Central Colombia. In appearance it is similar to the previous species. The IUCN has registered it as a rare species.

Old World or Thin-nosed Simian Primates (infraorder Catarrhina)

The huge geographical barrier of the Atlantic Ocean forms the dividing-line between two primate realms whose evolution began in the Paleocene age, about 60 million years ago. Fossil remains from this and the ensuing periods testify to the mutually isolated evolution of the two groups. There are naturally substantial gaps in this fossil record which for the time being have to be bridged with assumptions in order to obtain a comprehensive picture of primate evolution from the basic group of insectivores up to the great apes and ultimately to human beings. A fascinating aspect is the host of species and forms engendered by the evolutionary process that is governed by natural laws with their causes and effects. As the scientific name indicates, the Old World primates have noses with thin bridges. Other characteristic features are: a tail that is not prehensile, a longer auditory canal and also species that range in size from the dwarf guenons (about 1.3 kilos) to the powerful gorilla (up to about 250 kilos). They are classified into two superfamilies, the Cercopithecoidea and Hominoidea.

Cercopithecoid Monkeys (superfamily Cercopithecoidea)

The powerful and often jutting facial bone structure with the extremely well-developed canines closely resembles that of a dog. The superfamily includes monkeys with bodies measuring from 32 to 110 cm and tails that vary greatly in length, from short stumps to more than 100 cm. Some species have no tail at all. Other typical features are flat nails on all fingers and toes, various sized hard patches on their hindquarters (ischial callosities), movement on all fours, cheek pouches, gestation lasting from 165 to 240 days, daytime activity and a social way of life.

Family Cercopithecidae

The peoples of the ancient civilizations already took a lively interest in the South Asian and African species of this family. Numerous artistic representations testify to this. The ancient Egyptians regarded some species as sacred animals. Because they resembled men, physicians in Ancient Greece and in medieval Europe dissected these monkeys in order to find out more about the anatomy of the human body which they were not allowed to use for this purpose. These monkeys were displayed in the zoological gardens of the Egyptian Pharaohs and Chinese imperial dynasties and, down to the present day, the species of this family embody the quintessence of simian life. They are omnivorous to a more pronounced degree than other simian families and in accordance with their restless and mobile habits spend most of their lives eating. Over recent decades some startling facts have come to light about their life in the wild, providing new knowledge about their natural behavior and valuable biological information for zoological gardens as well as correcting outdated ideas. In particular we have learnt more about the sociological structure, organization and interplay of ecological factors affecting those species, such as the baboons which live in open country. The family is classified into eight genera.

Macaques (genus *Macaca*)

Their body length ranges between 38 and 76 cm; the tail is about 60 cm long; they weigh up to 13 kilos, and in a few cases males may weigh up to 16 kilos. With one exception—the magots—they are only found in South and East Asia. Thickset bodies with powerful limbs are typical features. Adult animals have large swellings on their hindquarters. Gestation lasts about 175 days; their reproductive organs are mature when they are about five years old. They live on the ground as well as on rocks and in trees. All macaques possess a very well-developed communication system composed of a wide range of facial expressions, gestures, postures, and sounds. The divergent forms that are found in the vast area of their distribution have resulted in a classification of six subgenera with twelve species.

Magot or **Barbary Ape** *(Macaca sylvana)*. The body measures 75 cm with a shoulder height of about 50 cm; elderly males have conspicuous bulges over their eyes. The long, thick, light to medium-brown fur makes the animal appear larger than it is; it has no tail. Magots may live up to 25 years. Their range extends in northwestern Africa from the Atlas Mountains to the coasts of Morocco and Algeria. The ancestors of the colony living on the Rock of Gibraltar were probably brought there around the first century A.D. Magots can therefore be regarded as the only members of their genus living outside Asia, the sole simian species found north of the Sahara and the only non-human primates in Europe living in the wild. These hardy monkeys live in troops numbering from 12 to 35 animals in wooded regions of the Atlas Mountains that are often snowclad for several months, in cedar woods and other areas with a sparse tree cover as well as in rocky regions covered with bushes and shrubs.

The social life of the magots is one of the most peaceable and placid in the primate world. Changes in the leadership of the herds are infrequent. The order of precedence determines

Magot or Barbary ape *(Macaca sylvana)*

priority in terms of food and water, comfortable resting-places and desirable partnerships. The segregation of the sexes is a distinct feature. Quite often the majority of males of all ages gather together in a confined space. On the other hand, the adolescent females remain with their mothers for a very long time forming regular maternal families. Like the hamadryas baboons, low-ranking adult males amongst the magots also use the children as a kind of protective buffer. A. Jolly (1975) describes how they carry the young around when these are just a few weeks old, caressing their bodies and presenting the babies to dominant males as a legitimation, even sitting down within reach of such a male with a baby between them—the polite response of the dominant male being to lift the baby's tail and kiss its hindquarters. In such cases semi-maternal behavior on the part of a young male serves as a kind of status symbol. Deag calls this behavior "social buffering." Current studies are aimed at determining whether the intensity of the male's affection for the children is governed by his progenitive powers. It is possible to produce proof of paternity as indicated by successful tests in Rheine Zoo (Federal Republic of Germany).

The Gibraltar magots are particularly celebrated. Officially imported for the first time by the garrison in 1856, they were afterward placed under British protection. The magots thrived and spread over the whole town during the following years. But they grew too mischievous, causing much damage. On one occasion during an official ceremony they stole the plumed hat of the British governor fleeing with it to the high walls of the fortress where they tried it on. This escapade marked the end of the patience and toleration with which they had hitherto been regarded and they were banished to a more isolated part of the Rock. But the tradition was still upheld, although in 1913 it was decreed that the number of the colony should not exceed 40; in 1910 there were more than 200 of them. The Gibraltar monkeys are under the administration of the British military authorities and the registered animals have their own "officer in charge of apes" and a per capita financial allotment for their food rations. But the colony has occasionally become reduced in numbers and more animals have had to be imported from North Africa.

Another successful settlement of these hardy monkeys in Europe deserves mention. In 1763 Count von Schlieffen brought a group of magots to his wooded park in Windhausen near Kassel. Apart from the natural forage that they found for themselves in the woods, they were given additional food. The animals sheltered from the rigors of the climate in some little huts and deeper cavities in the rocks. They acclimatized well and bred successfully so that the herd grew considerably. Twenty years later, after a dog with rabies had bitten and infected them, they all had to be shot. Today a memorial in Wildhausen Park still marks the spot where the 60 magots of this colony are buried.

The magot is often to be seen in zoological gardens and in temperate zones thrives in large open enclosures. In the Rheine Zoo, for instance, a group which now numbers 40 animals has been spending the winter in the open air without harm since 1954, providing opportunities for instructive field studies.

Lyssodes (subgenus *Lyssodes*)

The following two species present us with macaques that closely resemble the magots in appearance although their areas of distribution are about 12,000 kilometers apart. Particularly the forms that live in northern and mountainous regions have adapted themselves extremely well to the climatic conditions. Their heavy, thickset bodies, which appear even more massive because of their long-haired fur, are in themselves an adaptation to life on the ground. The tail has become a kind of rudimentary stump; it is naked and flattened, incapable of much movement, bent to the right and represents functionally a mere adjunct to the hindquarter patches—a sort of "cushion." In addition this anal region serves as a signal, proclaiming amicable and friendly intentions when displayed with wagging tail stump to another member of the species.

Stump-tailed Macaque *(Macaca arctoides)*. Originally named *Lyssodes speciosa*. It possesses a long mane on its shoulders and equally long sleek fur on its head. The face is red, sometimes patchy in places. When excited or hot the flush becomes deeper; when the animal is cold or out of sorts it assumes a blueish tinge. The young are still pale faced. The head appears very massive and the face large. Knowledge about these animals in the wild is still incomplete. While traveling over longer distances the small groups usually unite to form large herds numbering far more than a hundred animals. They cause substantial damage to the plantations in the valley regions. They are generally mountain-dwellers, distributed over wide zones of Southeast Asia from 33° northern latitude to the coasts of the Chinese Yellow and Eastern Seas via Vietnam, Thailand, Burma and Assam to eastern Tibet. The regionally varying forms are classified into five geographical subspecies which towards the northern regions have larger bodies and shaggier coats. Acclimatized to the cold, these monkeys sometimes wade through deep snow. Under these conditions their diet consists only of soft bark, buds, dried berries, and seeds. This animal is occasionally exhibited in zoos; both here and more especially in the larger primate research centers it reproduces successfully.

Japanese Macaque *(Macaca fuscata)*. This macaque, too, has a red face, although somewhat lighter in hue than the previously mentioned species. It has a body length of about 60 cm and the shoulder height is ca. 45 cm. The back is reddish brown to beige, while gray tints predominate on the underside. Its thick, long-haired, shaggy coat makes the animal appear fatter than it really is. They have survived to an age of 35 years in zoos. The border of their northern distribution is approximately 42° northern latitude. On the Japanese islands they are distributed only locally; there are none on Hokkaido. In winter the Japanese macaques have to be hardy enough to survive long periods of frost with thick snow and meager sources of sustenance. A large herd of macaques in the Jikogu Dani valley of Nagano Province takes baths during the winter in some hot volcanic springs. Since the animals are not in the least afraid of water they soak up to their necks in it, delousing themselves in the process while the young keep warm by splashing about and playing.

In adaptation to this unusual climate for monkeys the Japanese macaques have evolved a seasonal reproductive cycle lasting from November/December to March/April. The young born in the spring and summer have time to develop the necessary resistance to the rigors of the climate before the next winter begins.

In 1948 a group of Japanese scientists began an intensive study of the behavior of a group of macaques living on the islet of Koshima. By feeding them the scientists managed to make contact with the various groups, to identify individual members and establish their relationships, and thus became familiar with their social structure and could compare the groups. Some dramatic results have been achieved since then. A single group, which may number five, but can also run into several hundreds—the average is from 60 to 120 animals—, forms a well-knit community. Its social pattern, "organized" in terms of order of precedence and age groups, is well coordinated and evenly balanced even when its territory is relatively confined. In conformity with the available food supply and size of the group, it lays claim to territory ranging in area from about 2 to 15 square kilometers which the animals constantly comb in search of fodder. They find safe sleeping-places in trees in the woods or on virtually inaccessible rocks. On the move the group is headed by a troop of young males, followed by the females with children and adolescents and some of the older males, while the rear is brought up by another troop of young adult males. A similarly well-defined order can also be observed at assembly and feeding places. Relations within the community assume two basic forms: heterogeneous bonds which include mating relationships and those between mother and child; and homogeneous relations between similar animals in which predominance and rank form the major factors. Until they are about ten months old the children are taught by their mothers. Afterward they play more and more with companions of their own sex. After about two years—during which period the young stay with their mothers in the heart of the group—the adolescent males join their peers on its fringe. The young females, on the other hand, remain within the maternal circle. The males' social advance to leadership is determined to a lesser degree by grim battles; play, hunting and wrestling are more important and enable the young males of a specific age group to become familiar with each other and to assess a rival's potentialities. After a great deal of shuffling their social status finally emerges in early manhood. But other factors, such as the social status of the mother, and perhaps of the father too, genetic endowments and the intensity and comprehensiveness of his education, also play a part in a male's advance to a dominant position. The females can reproduce when they are four years old, the males not until they are five. The high-ranking males naturally enjoy priority in the choice of a mate. But if there are more sexually receptive females than the leading males can pair with, males from the outer circle may choose a mate without incurring the wrath of a dominant member of the group. In contrast to the males' well-defined class and social structure, the social status of the females is subject to certain fluctuations usually caused by individual conflicts and quarrels. The daughters of a high-ranking female occupy a leading position because they remain under their mother's influence and are thus dependent on her for a long time. Amongst the Japanese macaques sexual activity is not the decisive factor for social links within the group since social contact is permanent and not restricted to the previously mentioned mating season. The fact should naturally not be overlooked that there is occasional copulation with ritual significance—for instance for the settlement of individual conflicts. To a greater extent than was previously assumed, the bonds between mother and child and—as we now know—between father and child are a cementing factor. Since closer bonds also exist between a male and his respective mate, as reflected in the grooming procedure, these definite "threesome" contacts may be interpreted as the beginnings of a family. Paternal behavior on the part of high-ranking males takes the form of looking after the one-year-old children when their mothers have a new baby that absorbs all their time and attention. This paternal care is primarily concentrated on a certain child. Sometimes loners—always males—are encountered. There are various possible reasons why they live outside the community—change of group, migration, age, discrepancies in the order of precedence, loss of rank—but asocial motivations are unlikely.

---

65 Wanderoo or **Lion-tailed Macaque**
(*Macaca silenus*).
In captivity lion-tailed macaques are hardy animals with a good life expectancy and are not particularly difficult to breed.
(Liberec Zoological Gardens)

66 Stump-tailed Macaque *(Macaca arctoides)*.
These macaques are hardy mountain-
dwellers, accustomed to low temperatures.
(Prague Zoological Gardens)

67 Japanese Macaque *(Macaca fuscata)*.
These monkeys play an important role in
Japanese mythology, history and folk tales.
Everyone is familiar with the three wise
monkeys that embody the Buddhist principle:
"Hear no evil, speak no evil, see no evil."
(Stuttgart Zoological Gardens)

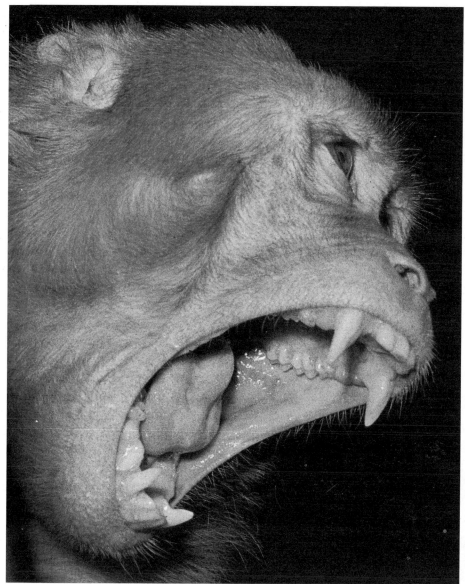

68–71 Rhesus Monkeys *(Macaca mulatta).*
Above left: Older males in particular have very long canine teeth. (Magdeburg Zoological Gardens)
Above right: Left alone, the young rhesus monkey cries bitterly for its mother. (Ostrava Zoological Gardens)
Below: Liberec Zoological Gardens

72 Wanderoo or **Lion-tailed Macaque**
(Macaca silenus)
(Liberec Zoological Gardens)

73 Celebes Crested Macaque
(Cynopithecus niger).
Some tribes on Sulawesi believe they are
descended from the crested macaque and respect
and venerate this species as their ancestors.
In other regions these macaques are hunted
for their flesh. (Hanover Zoological Gardens)

74 Crab-eating Macaques *(Macaca irus).*
In contrast to its mother, the young animal
is dark brown. (Sacred Monkey Forest on Bali)

75/76 Wanderoos or **Lion-tailed Macaques**
(Macaca silenus).
A long gray ruff-like mane whose fur curves
forward is a characteristic feature of the
lion-tailed monkey.
Left: Hoyerswerda Wild-life Park
Right: West Berlin Zoological Gardens

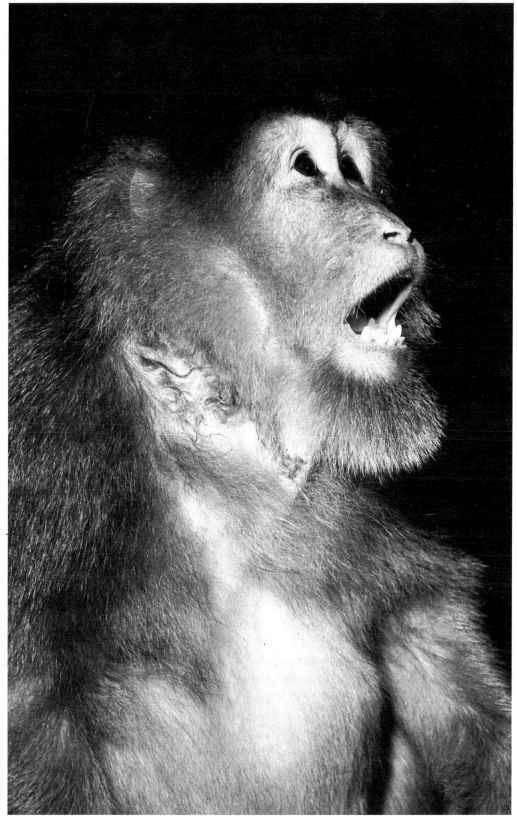

77 Japanese Macaque *(Macaca fuscata)*.
The thick fur on the head protects the ears
and face from the cold. (Prague Zoological
Gardens)

78 Celebes Crested Macaque *(Cynopithecus niger)*.
The older animals' upstanding crest
of fur falls toward the back of the head.
(Hanover Zoological Gardens)

Overleaf:

79 Yellow Baboon *(Papio cynocephalus)*.
Like most baboons, these animals are very
adaptable, and often cause considerable damage
to plantations. (East Africa)

80 Young Yellow Baboon *(Papio cynocephalus)*.
(Hanover Zoological Gardens)

81/82 Yellow Baboons (*Papio cynocephalus*)
in the East African scrubland. The male's
threatening yawn warns the photographer not
to come any closer. The coat is relatively
thick and long-haired.

83 Sacred or **Hamadryas Baboon** (*Papio hamadryas*).
At play or when speed is called for the
young baboons often climb onto the backs of
older animals. (Cologne Zoological Gardens)

Overleaf:

84/85 Sacred or **Hamadryas Baboons**
(*Papio hamadryas*).
According to available records a total of 290
hamadryas baboons were born and reared
in 66 zoos in 1978. As the animals grow older
the conspicuous mane which covers the head
like a white fur hood turns a silvery gray
round the chest. (Cologne Zoological Gardens)

86–88 Sacred or **Hamadryas Baboons**
(Papio hamadryas).
Above: When the male threatens with a yawn
his long dagger-like canine teeth are fully
exposed. If a dominant male happens to lose
these formidable weapons his days of leadership
are over. (Liberec Zoological Gardens)
Above right: Apart from his size and thickset appearance the male hamadryas baboon
can be recognized by his magnificent
long-haired gray mane that covers the
shoulders, chest and top part of the back.
(Prague Zoological Gardens)
Right: The young are kept under strict control
and are not allowed to stray beyond their
mother's reach. (Cologne Zoological Gardens)

89–91 Sacred or **Hamadryas Baboons**
(Papio hamadryas).
Above left: Grooming is a social activity
in which the familiar keeper is included.
He hence occupies a high-ranking position.
(Cologne Zoological Gardens)
Above: Liberec Zoological Gardens
Below left: Young baboons at play. They
have a pronounced activity urge.
(Hanover Zoological Gardens)

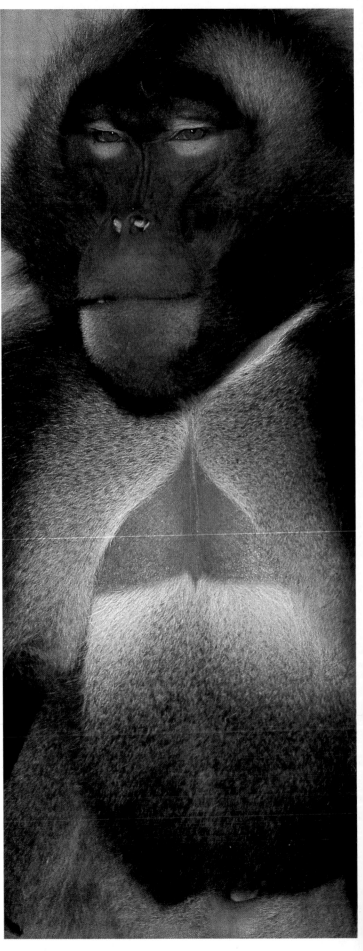

92 Mandrill *(Mandrillus sphinx)*.
The conspicuous fur round the face and the cheek
markings help to intimidate rivals.
(Dvur Kralove Zoological Gardens)

93 Gelada *(Theropithecus gelada)*.
Both sexes are furnished with a patch of
nakcd red skin bisected by a furrowed line
on the chest. (Duisburg Zoological Gardens)

94 Mandrill *(Mandrillus sphinx)*.
Female mandrills are much smaller, slimmer
and less garishly colored than the males.
(Dvur Kralove Zoological Gardens)

95 Mandrills *(Mandrillus sphinx)*.
When their mother is absent the two young
mandrills seek comfort by clinging to each
other. The typical scored ridges on each side
of the nose are already apparent.
(Prague Zoological Gardens)

96 Drill *(Mandrillus leucophaeus)*.
Compared with the mandrill, the drill's
coloring is less startling, only its chin being
bright red. (Stuttgart Zoological Gardens)

97 Mandrill *(Mandrillus sphinx)*.
Display behavior on the part of a powerful
male mandrill presenting himself to zoo
visitors. (Duisburg Zoological Gardens).

98–100 Geladas (*Theropithecus gelada*).
Above left: Threatening behavior includes baring the upper lip thus effectively displaying its bright pink inner surface. The raised eyebrows reveal the conspicuously light eyelids. (Stuttgart Zoological Gardens)
Above right: When on the look-out, reconnoitering potential sources of danger or when displaying, the geladas often stand erect. (Stuttgart Zoological Gardens)
Below: When resting during the day the geladas devote much time to mutual grooming whereby the dominant males take priority. (Stuttgart Zoological Gardens)

101 White-nosed Patas Monkey
(Erythrocebus patas pyrrhonotus).
The young males reach sexual maturity when
they are 3.5 to 4 years old. With fierce
onslaughts the dominant male then expels
these potential rivals from the group.
(Dvur Kralove Zoological Gardens)

102 Vervet Monkey
(Cercopithecus aethiops aethiops).
Guenons are generally polygamous and mating
partnerships are only of a brief duration.
As a rule they are not so easy to breed as
macaques and baboons.
(Dvur Kralove Zoological Gardens)

103 Gray-cheeked Mangabey
(Cercocebus albigena).
These mangabeys are more timid and reserved
and descend to the ground less frequently than
the other species. (Hanover Zoological Gardens)

104 Black Mangabey *(Cercocebus aterrimus).*
Mangabeys are usually found in
smaller troops numbering from 8 to
15 animals, and sometimes only in pairs.
(Prague Zoological Gardens)

105 Greater White-nosed Guenon
(Cercopithecus nictitans).
The white spot on the nose may vary
in shape. The white-nosed monkeys are
today still a fairly common species.
(Duisburg Zoological Gardens)

106 *Cercocebus torquatus atys.*
The white eyelids form a distinct contrast
to the dark face. They gleam like little
signal lamps when the animal blinks and
serve as a means of visual communication
as well as indicating its position in the
dim light of the tropical forest.
(Thuringian Wild-life Park Erfurt)

107 Red-nosed Guenon
(Cercopithecus cephus erythrotis).
Opinions differ as to whether this guenon
belongs to a subspecies of the moustached
monkey or is a distinct species.
(Duisburg Zoological Gardens)

108–110 Vervet Monkeys
(Cercopithecus aethiops sabaeus).
Left: About 20 different subspecies have
been identified. They vary mainly in the head
markings, fur coloring, size and in the color
of the male's scrotum. (Prague Zoological
Gardens)
Right: During the clinging age the baby's
tail is slightly prehensile and is wrapped
round the mother's tail, abdomen or flanks.
The birth of twins is uncommon. Adolescent
animals still maintain close contact to the
mother even when she has a new baby.
(Prague Zoological Gardens)

111 Moustached Monkey
(Cercopithecus mitis kibonotenensis).
This subspecies of the diademed guenon has a blueish gray coat and a flowing beard. In East Africa they occur in concentrated zones near Mt. Kilimanjaro and Mt. Meru. Like all diademed guenons they are of a quiet and reserved disposition. (Hanover Zoological Gardens)

112 White-throated Guenon
(Cercopithecus mitis albogularis).
The animals with lighter head coloring and white throats are grouped together as white-throated guenons to distinguish them from the "true" diademed guenons. (Duisburg Zoological Gardens)

113 Diana Monkey (*Cercopithecus diana*).
Agricultural encroachments and extensive timber-felling are driving the Diana
monkeys out of their natural habitat.
(Liberec Zoological Gardens)

114 Mona Monkey (*Cercopithecus mona* subspec.)
Mona monkeys have a more independent and
reserved disposition than, for instance,
the vervet monkeys. (Liberec Zoological
Gardens)

115 Roloway Monkey (*Cercopithecus diana
roloway).*
Its expression clearly reflects its mood.
(Duisburg Zoological Gardens)

116/117 De Brazza's Monkeys
(Cercopithecus neglectus)
live in small groups. The more powerful and
brighter colored male only tolerates a few females,
sometimes just one or two, in his immediate
proximity. (Liberec Zoological Gardens)

118 Hamlyn's Monkeys or **Owl-faced Monkeys**
(Cercopithecus hamlyni).
Hamlyn's monkeys live in the depths of the
tropical forest. Their movements appear elegant
and graceful. (Leipzig Zoological Gardens)

Some very interesting observations about the emergence of new habits were made by the Japanese zoologist M. Kawai (1965) who studied a macaque group on the islet of Koshima. It all began when a young female whom the observers had named Imo picked up a sweet potato covered with sand, dipped it into the water and washed it before she ate it. About a month later one of her companions began to do the same and some months later Imo's mother followed suit. In the course of daily contacts between the mothers, young females, children and their peers 15 animals had already adopted this habit from 1953 to 1957. Urged by their natural curiosity, it was chiefly young animals aged between one and three years who took advantage of the new discovery. On the other hand, none of the adult males ever learned to do so, possibly because the novelty was introduced by the circle of young females and children. The enterprise and curiosity of the younger members of the community is offset by the experience, caution and reservation of the older adult animals. After five years all the newborn had already learned this new behavior from their mothers and after ten years for more than half the 59 members of the group this habit had become universal. To begin with the macaques only washed their sweet potatoes in freshwater pools, until some of them while traveling and bathing along the coast began to wash the potatoes in seawater, obviously relishing the salty taste. This became a general habit and the macaques started to dip their sweet potatoes into the seawater while eating them. It was observed that divergent genetic endowments and abilities played a part in this procedure.

In 1956 the four-year-old Imo made another innovation. The macaques were sometimes fed with grain and originally had to pick out grain for grain from the sand. But Imo grabbed a handful of grain, sand and all, and threw it into the water. The sand soon dropped to the bottom but the grain floated and she skimmed it off with her hand. By 1962 19 animals had adopted this habit. Imo's brothers and sisters learned better and more rapidly than the offspring of another female macaque.

The Japanese macaques played a prominent role long ago in the mythology, literature and art of the Japanese people. The three wise monkeys, contrary to general belief, did not originate in India but in Japan where they embody a principle of the Buddhist religion: "See no evil—hear no evil—speak no evil." The macaques are under strict protection in Japan. Because they can stand low temperatures they are fairly common in zoological gardens. There are two subspecies of the Japanese macaques:

Macaca fuscata fuscata, found on all the larger Japanese islands except Hokkaido; relatively common near Kyoto; *Macaca fuscata yakui*, still fairly common on the smaller island of Yaku-Shima to the south of Kyushu.

Rhesus Monkeys (subgenus *Rhesus*)

As compared to the powerful body of the male which can assume massive proportions in old age, the females appear much slimmer and smaller. The fur is light to medium brown with a reddish brown or light olive-green tint; the face varies between pale grayish-brown and pink. The tail is of medium length—from 30 to 35 cm. Particularly in the case of adult males the scrotum and penis are a startling red. Rhesus monkeys are distributed over large areas of South and Southeast Asia. There are three species.

Rhesus Monkey (*Macaca mulatta*). One of the commonest and best known monkeys, already familiar to Europeans in bygone centuries as a performing animal in traveling menageries, at annual fairs, etc. It is found all over the Indian subcontinent (with the exception of southern Pakistan), eastward to the shores of the Eastern Sea, in the north nearly as far as the Hwang Ho and in the southeast to the borders of Laos, Thailand and Burma. Rhesus monkeys are frugal in their demands, adaptable and hardy by nature. Today only a minority of them inhabit forests and rocky country, the majority being found in the vicinity of cultivated land, settlements and even in the heart of the cities. They steal from homes and cars if the windows are left open. Even though rhesus monkeys are not regarded as directly sacred, the Hindus coexist peacefully with them in accordance with their religious views about animals: the damage they cause is tolerated with long-suffering patience. Earlier attempts to transfer this macaque from densely populated areas to more suitable zones were usually defeated as a result of protests from the pious Hindu population. Even parliamentary debates did nothing to change the situation. According to a territorial census carried out by C. H. Southwick (1965) in Uttar Pradesh (North India), 46 percent of these monkeys live in villages, 30 percent in towns with a scattering in areas near roads and temples, and only 12 percent in wooded zones. According to the climatic conditions under which they live the monkeys have either a thin sleek silky coat or thick woolly fur. The rare gold rhesus monkey with its pale golden fur forms a variation in the color scheme.

119 Hamlyn's Monkey or **Owl-faced Monkey**
(*Cercopithecus hamlyni*).
An interesting feature is the color scheme which, like the common hamster or young panda, has a "back-to-front" pattern: the dorsal side is gray-brown-green, the ventral side, legs and arms are black.
(Leipzig Zoological Gardens)

The rhesus monkey is an able swimmer and diver. Indeed these monkeys prefer zones where there are stetches of water. They often spend hours in various sized groups searching for anything edible, their diet consisting of various kinds of grain, seeds, fruit, berries, roots, bulbs, tubers, buds, blossoms and a host of small animals. Although mothers with newborn babies can be encountered in the towns at all seasons, the majority of births occur during the spring and early summer. At this time and for some months afterward, and especially when living at a higher altitude, the female has no sexually receptive period and therefore does not conceive. According to Michael and Kaverne (1968) the female's vaginal odor is the main attraction for the male rhesus, which signifies stimulation through estrogen, since the female rhesus has only slight sexual swellings on her outer genitals.

The mother and child of lower social status are not always treated as indulgently as those of higher rank. Young females begin to mind the babies temporarily at an early age. And according to Spencer-Booth (1968) females who have not yet given birth to offspring are more inclined to behave in this manner than successful mothers. Although the children are carefully looked after as a general rule, they soon become familiar with the rough and tumble of group life. Siblings also frequently remain together when they reach adulthood. Amongst the rhesus monkeys, too, the males often make use of the children in their bids for status. Those of them who feel an urge to get into the center of the group take care of the children of dominant females and are therefore tolerated in the inner circle in the proximity of the mothers. These macaques may live for over 30 years. A rhesus monkey with no social contact soon shows signs of disturbed behavior and makes a woebegone impression, similar to the baboons in such circumstances.

In order to survive on the ground these monkeys had to adapt themselves to the far greater dangers lurking there as compared to an arboreal life. To be successful they had to develop their combative and defensive powers to an extent where self-defence verges on aggressiveness. The highly developed social system and strict order of precedence engendered a "leadership" which ensures unity and uniformity in both behavior and action, augments their fighting power in defending themselves, and thus demonstrates the strength of the community. If, for instance, a member of a group in a zoo is intimidated or harassed by a keeper, the whole troop menaces the intruder with angry gestures and loud hoarse cries; they not infrequently resort to violence too. Comparing the behavior of rhesus monkeys in the forests and in urban areas, D. S. Singh (1968) observed that the latter were more active, wary, manipulative, alert, clever, indeed more artful and impudent. An interesting feature is the ritualized behavior between superior and inferior animals. To begin with, the animal of lower rank is threatened with gestures such as an opened mouth, bared teeth, flattened

ears, etc. which originally signified an imminent attack. If the inferior is intimidated by the pugnacious behavior of the superior or has no intention of challenging the latter's authority, he adopts an attitude of submission which has an appeasing effect because it is that of a female during copulation. The dominant male even mounts the inferior as if for mating. Through this more or less symbolic act, which has nothing to do with sexuality, the submissive male placates or mollifies his potential attacker. The dominant status of the other male is recognized, there is no serious struggle and injuries are avoided. By overcoming internal conflicts in such a harmless or amicable manner the group husbands its powers of self-defence against external enemies. Like mutual grooming and the communal care for children this ritualized behavior helps to preserve the texture of a well-knit community.

Especially in captivity rhesus monkeys have caused astonishment with their amazing intelligence—which of course varies from individual to individual. Because they are so easy to train groups of them can be taught difficult tricks which they sometimes perform in circuses. In India exhibitors have achieved some remarkable results with rhesus monkeys. A female rhesus at the Münster Zoological Institute learned to use colored rings as counters of different value in exchange for titbits rated according to their popularity with her. Feats of this nature are otherwise only known to have been achieved by chimpanzees or orang utans. A parallel achievement under far more hazardous conditions was the space flight of the female rhesus "Lizzie" in a Mercury capsule in 1949. In the course of an intensive training program she had been taught to operate a complicated system of levers, buttons and keys in the correct sequence in response to certain signals, release controls and other stimuli. In Wisconsin Regional Primate Center extensive studies on the way rhesus children behave to each other were conducted by H. F. Harlow (1959, 1960). He came to the conclusion that primates express their affection in five distinct forms each of which can be incorporated into its own "affection system"—for instance, love for the mother, the affection of an adult male for the offspring, comradeship between animals of the same age group or between companions, etc. Nervous primate children or those under stress seek maternal protection when alarmed or in danger. Even contact with a dummy made of soft material is sufficient to allay fear. In the course of later experiments in Davis (California), W. A. Mason and M. D. Kinney (1961, 1968) demonstrated the great importance of physical contact with a living animal for the emotional social well-being of a primate child. A good-tempered dog, for instance, can satisfactorily replace the mother for a young rhesus monkey irrespective of whether it was used to being with its own mother or had a foster-mother. Previous views that the affection of an infant for its mother is primarily aroused because she is the giver of food, who stills the pangs of hunger, have not been endorsed. In the

Although the wired puppet offers sustenance to the young rhesus monkey *(Macaca mulatta)*, it clings to the upholstered dummy that gives it warmth and comfort.

opinion of H.F. Harlow the father, too, can arouse and keep the affection of the child to the same extent as the mother. The instinctive urge for a satisfying natural outlet for the clinging reflex, for caressing touches and comforting emotional warmth, as well as the feeling of security, are the principal factors. To be a good mother, a female monkey must herself experience the best possible maternal care. Females reared in isolation without this comfort do not lavish nearly as much affection and care on their own children as female monkeys living in the wild. Comfort and security are therefore the major prerequisites for the monkey's healthy development.

In medical research, too, important advances have been made with the aid of rhesus monkeys (see p. 17). Because so many rhesus monkeys have been used during recent decades for experimental purposes in medical research and the pharmaceutical industry, the herds in several regions, in Bengal for instance, have been ominously depleted. For this reason a breeding farm for monkeys was established near Bombay where the animals grow up under maximum natural conditions. A colony of rhesus monkeys, primarily intended for behavioral studies, live on the little island of Cayo Santiago near Puerto Rico in the Caribbean. One of the observers, C.B. Koford (1968), conducted interesting studies about their population dynamics and kinship links.

Like baboons, the rhesus monkeys react aggressively when attacked by enemies. They are still fairly common in zoological gardens and generally breed well.

Over the vast area of their distribution geographical subspecies have emerged, which, because of the great adaptability of the rhesus monkey, cannot always be exactly defined. Apart from the nominate form *Macaca mulatta mulatta*, I.T. Sanderson (1957) classified the subspecies *Macaca mulatta vestita*, occurring in the Tibetan-Indian borderland; like the two following forms it closely resembles the Assam rhesus *Macaca mulatta villosa* inhabiting the Upper Punjab, Kashmir, and northern Uttar Pradesh, and *Macaca mulatta memakoni* of northern Pakistan and eastern Afghanistan. The largest and heaviest forms inhabit the western Himalayas to an altitude of 2,400 meters.

Assam Rhesus Monkey *(Macaca assamensis)*. With a body length of more than 60 cm, a tail that is about 20 cm long and a weight of up to 12 kilos, this species is larger, stronger and also more thickly haired than the true rhesus. The fur varies in color from brown to light grayish-brown, the face is pale pink. Found in altitudes up to 2,000 meters, it lives in large groups which often split up into smaller troops numbering from 8 to 20 animals. In the Sikkim and Darjeeling regions they descend to an altitude of 600 meters when the weather is cold. On my way to Darjeeling in 1960 I saw in the upland forests a herd of these monkeys crossing the Peshoke Road. First of all a strong young adult male reconnoitered on the edge of the little-frequented road. Then he was rapidly followed almost in rank and file by 14 animals, as far as I could see mainly females, and including three young monkeys. At the rear trotted the powerful elderly dominant male who also paused a moment at the edge of the road to reconnoiter.

According to I. T. Sanderson they even, to a certain extent, defend their territory in the mountains and hills, rolling stones down on intruders which is virtually equivalent to an aimed throw. Their range extends in the Himalayas from Kashmir eastwards via Nepal, Bhutan, the mountain ranges of Assam, the northern part of the Sunderbans, Upper Burma, Yunnan, Kwang-si to northern Vietnam. In Sikkim and in the Darjeeling region these macaques are hunted for their flesh by the Lepchas who believe it has curative properties. In zoological gardens these macaques are not very common. Two subspecies are known:
Macaca assamensis assamensis, occurring in Assam, via Upper Burma to northern Vietnam; *Macaca assamensis pelops,* to be found in Bhutan in the higher altitudes via Sikkim, Nepal to Kashmir.

Formosa Macaque or **Formosa Rhesus** *(Macaca cyclopis).* This species of the subgenus *Rhesus* was originally distributed over the mountainous region of Taiwan where there was little tree cover and they had to travel long distances to find sufficient sustenance. When danger threatened the animals found a safe refuge in barely accessible rock cavities. Under the cover of darkness they often raided the peasants' crops. Today the monkeys have been largely ousted, indubitably as a result of Taiwan's dense population. They now live on the rocky coastal strips and have adapted their diet by and large to what the beaches have to offer to them. Any details about their habits are not known.

Subgenus *Silenus*

Only two, superficially very dissimilar species, are included in this classification. There are similarities in the fur on the head, especially in the bearded parts, as well as in the shape of the tail and the female's genital swellings. Found over large parts of South and Southeast Asia.

Wanderoo or **Lion-tailed Macaque** *(Macaca silenus).* The body length ranges between 50 and 60 cm, the tail measures from 25 to 40 cm and the males may weigh up to 10 kilos; the females are much smaller. A conspicuous feature, especially among the males, is the huge, mane-like gray halo of fur that frames the face very decoratively. The black, glossy coat is sleek, and the tail has a tuft at its end, from which this species derives its name. As in the case of the rhesus monkeys, the order of precedence is enforced by a powerful dominant male. According to Indian sources, lion-tailed macaques are said to behave in a very aggressive way in their territory, even occasionally attacking human beings, and the powerful canines of the adult males are certainly to be treated with respect.

In troops numbering from 12 to 20 animals and in small families the lion-tailed macaques roam the mountain forests of the Western Ghats at an altitude of between 600 to 2,000 meters, from Goa southwards to Cape Comorin and the southwestern part of Tamil Nadu. Although they mainly live on the ground, they are also found more frequently than rhesus monkeys on the coniferous trees where they are hard to discern. They are also less carnivorous in their diet. Newborn animals are usually seen in South India in September.

The IUCN has registered them as an endangered species that is threatened with extinction. In the past they were frequently exhibited in zoos. There are no known subspecies.

Pig-tailed Macaque *(Macaca nemestrina).* With a body length of up to 70 cm, shoulder height of up to 50 cm and weighing up to 15 kilos, this is one of the largest macaques. The thin tail, measuring only from 15 to 20 cm, is carried curled upward in a manner somewhat similar to that of the domestic pig. Despite the comparatively long legs, its appearance gives the impression of strength and heaviness. It has a sleek and gleaming brownish to yellowish-green coat. There is a patch of darker fur on the top of its head and its eyelids are whitish. The pig-tailed macaques are distributed over a fairly large area: from Assam, via Burma, Thailand, the Malay Peninsula, Sumatera to Kalimantan. They live in groups averaging about 39 animals, mainly on the ground in the forest or near its fringe. Strong old males may grow very aggressive. By nature the pig-tailed macaques are more pugnacious than the rhesus monkeys. I. S. Bernstein (1969) observed a male engaged in battle with the dominant animal which died of the injuries he inflicted. Two months later the same animal attacked the male who had been respected as second-in-command until then, won the battle, killed the highest-ranking female and became the dominant male. This violent seizure of power can of course be engendered or promoted by a too dense population. Similar to other macaques, status in the group determines priority in choosing a mate, in access to food and the use of threatening behavior. The pig-tailed macaques have a special welcoming gesture for their friends: lifting their heads and raising their eyebrows they push out their lips to form a broad "pout." They even return this greeting when, for instance, a keeper with whom they are familiar imitates the gesture. In zoos it is fairly common to find them in smaller groups. Several subspecies are known:
Macaca nemestrina nemestrina, the largest form. Southern Thailand, Malay Peninsula, Sumatera; *Macaca nemestrina leo-*

nina, northern Thailand, southern Burma to Tenasserim; *Macaca nemestrina blythi,* found in northern Burma as far as Manipur and Nagaland. The two forms in Burma have a longer coat, the tail is very hairy and has a tuft at the end. *Macaca nemestrina andamanensis,* found on the Andaman Islands; *Macaca nemestrina pagensis,* Mentawai Islands west of Sumatera.

Bonnet Monkeys (subgenus *Zati*)

They belong to the smaller to medium-size macaques. The body is slimmer, the tail longer than in the case of rhesus monkeys. They have long fur on the head and a naked face. They represent the rhesus monkey in South India, approximately south of a line from Bombay eastward to Rajamundry (mouth of the Godevari).

Indian Bonnet Monkey *(Macaca radiata).* Known as "bandar" in Hindi this macaque has a body length of about 50 to 55 cm, the tail being longer (about 60 cm). The males weigh from 6 to 9 kilos, the females 3 or 4 kilos. The brow appears very high and is naked. The long dark hair on the head twirls outward. The short thick coat varies from medium to light brown in color. The bonnet monkeys are found from the previously mentioned northern border to Cape Comorin, in forests and villages, in the plains as well as on the hills and uplands, mostly in troops numbering from 20 to 30 animals. An arboreal life suits them better than it does other macaques. Their diet is almost indistinguishable from that of the rhesus monkeys. In contrast to the shy bonnet monkeys that inhabit the forests, those which live in the towns and villages, where they are universally tolerated by the Hindus, have lost all fear of human beings and even on very busy streets do not avoid contact with them. They usually find sleeping-places on roofs, in temples and so on. They possess a highly organized social order. Their social life is supervised by an inner circle of dominant males who, when necessary, combine to enforce law and order and to defend the group. Mating takes place all the year round whereby these activities reach a peak in October/November resulting in more births from February to April. Sexual maturity begins at from three to four years although the males do not attain their ultimate social status until two or three years later. It is estimated that they live to an age of from 12 to 15 years in captivity, and in a few cases up to 30 years (S. H. Prater 1971). They are common in zoos.

Ceylon Toque Monkey *(Macaca sinica).* Smaller body proportions than its cousin on the mainland. The longer mane on its head appears overgrown and hence untidy. The coat is of an almost uniform brown color. Although shy and easily alarmed, its almost insatiable curiosity keeps it constantly active and on the move. These macaques are not so common in zoos.

I. T. Sanderson (1957), basing his observations on the altitude at which they are found and the color of their coats, distinguishes three subspecies—a plains, uplands and mountain form.

Crab-eating Macaques (subgenus *Cynomolgus*)

They resemble the bonnet monkeys. Widely distributed, with only slightly differing varieties, they are grouped together as one species.

Crab-eating Macaque *(Macaca irus).* The body length is about 60 cm and the muscular tail is ca. 55 cm long. The upper side of the coat is gray brown with olive-green tints; the underside is grayish white. The face is of a blueish gray color, with a lighter whitish patch between the eyes. Both sexes are furnished with a short beard. These macaques inhabit a very extensive area of Southeast Asia from Burma via Thailand, Laos, Vietnam, Kampuchea, the Malay Peninsula, Sumatera, Java, Bali, Kalimantan as far as the Philippines, being chiefly found in wooded coastal regions, in mangrove swamps and on the banks of rivers and lakes from the low-lying plains to an altitude of 1,200 meters. In groups numbering from 14 to 70 animals—about 30 on average—they move about the flat shores and beaches and are able climbers, swimmers and divers. At high tide the various groups wait on the trees that afford a good view of the seashore. The best vantage points are energetically defended against neighboring groups and intruders. At low tide the animals clamber down from the trees and search nimbly and patiently for the prey they have spotted—fish, crabs and other crustaceans. But they can also exist on a vegetarian diet, at least as supplementary fare, when little else is available.

Their social behavior is more or less similar to that of the other macaques. Mutual grooming is accompanied by clickings and smackings of the tongue which have both functional and ritual significance. The female's outer genitals are extremely swollen and red when she is sexually receptive (estrus). This occurs at some time approximately between the 6th and 20th day after menstruation which lasts three to five days, the whole cycle taking about 28 days. The female with the biggest swellings stays in the proximity of the male, clearly soliciting his attention by repeatedly backing up to him, raising her hindquarters, sinking her head and often looking behind her. This period extends over about three days in the course of which several copulations take place daily. Gestation lasts about 164 days. The young are suckled for about 18 months. Sexual maturity is attained at ca. four years, somewhat later for the males. Only in exceptional cases do they live longer than 20 years.

Some peoples regard them as sacred animals. The Hindus among the Balinese still sacrifice to them in the form of food offerings. On Java too there are still monkey temples, partly as tourist attractions. In the USA large numbers of them are used for experimental purposes. A classification of these macaques into subspecies is extremely difficult. Although there are substantial regional differences in size and color, identical forms are found even in areas that are far apart.

I. T. Sanderson (1957) speaks of 21 subspecies, of which only one is mentioned here: *Macaca irus philippinensis*, the northeasternmost form.

Celebes Macaques (subgenus *Gymnopyga*)

The easternmost macaques, they resemble the magots in their type of body. All forms are classified as belonging to one species.

Moor Macaque *(Macaca maura)*. As the name indicates, this species has a completely black coat with thick fur. The facial bones are angular, the rudimentary tail resembles a button-like bump. Despite its thickset body the moor macaque, as a forest-dweller, is an adept climber and leaper. In its range it seeks the proximity of waterways around which groups of from 15 to 20 animals can find their food, hardly ever having to leave the forest. In addition to vegetarian fare it eats many kinds of small animals. The social order is upheld by a strong and older dominant male, whereby the macaque structure is the general rule. In some places the moor macaques were, and still are, regarded as sacred, while in others they are hunted for their flesh. Nowadays these macaques are no longer so common in zoos.

The busy trade in animals on Sulawesi has greatly upset the original geographical distribution of the various forms. According to Elliot there are three subspecies: *Macaca maura maura*, southwest peninsula of Sulawesi; *Macaca maura tonkeanus*, eastern peninsula, offshore Togian Islands; *Macaca maura ochreata*, silvery gray extremities and grayish fur on the sides of the head and ridges over the eyebrows; hindquarter patches deeper red in color. Southeastern peninsula, and on the islands of Muna, Butung and Wowoni to the south.

Crested Macaques (genus *Cynopithecus*)

Celebes Crested Macaque *(Cynopithecus niger)*. Although closely related to the macaques it is classified as a separate genus because of its mandrill-like cheek bones. Its body measures just over 60 cm while the very short tail is only a few centimeters long. It belongs to a somewhat lighter and long-legged simian type. The coat is jet black with an occasional dark gray tinge. The anterior skull with the jaws juts forward like that of

the baboons. The ridges over the eyes are very pronounced and naked. A typical feature is the bristly crest of hair on its head. There is a regrettable dearth of field studies about its habits. It is omnivorous, but prefers a vegetarian diet. In the mangrove swamps where this forest-dweller is often found, it is said to hunt various small animals too. It lives in groups of various size which are ruled by a strong male. But they are also found in pairs. Mutual grooming is a social activity which ignores status in the hierarchy. These macaques can be very aggressive. They utter loud barking sounds when fighting, threatening, warning or warding off attacks. According to observations made in San Diego Zoo, pregnancy lasts seven months. This macaque inhabits the northern peninsula of Sulawesi. In bygone days, but hardly ever nowadays, it was regarded as a sacred animal in some places, usually in connection with ancestor worship.

It is somewhat rare in European primate collections but fairly common in zoos in other parts of the world.

Baboons (genus *Papio*)

The baboons are the most typical representatives of the Cercopithecoidea. They are sturdily built, usually thickset, and only excelled by the apes in size and strength. Because of the jutting regions round the jaws and muzzle, the head appears elongated, heavy and bulky. The little deep-set eyes are placed relatively close together and protected by massive eyebrow ridges. Living on the ground, the baboons possess powerful limbs of equal length, short-fingered hands and feet with tough soles. Males and females often differ greatly in size. The body length ranges between 50 and 114 cm, the tail measures, according to species, between 5 and 70 cm, the shoulder heigth is from 45 to 60 cm, and the weight may vary between 14 and 54 kilos. The very powerful jaws are dominated, particularly in the case of the male, by dagger-like canines. Having gobbled up food hastily, the baboons can then store large amounts of it in their cheek pouches. The color of the coat ranges, according to species, from brown to gray, often with olive-green tints. Some of the males are furnished with a mane or a cloak of fur on their shoulder. The startling red anal and genital region, which in the case of the females is greatly swollen during the period of sexual receptivity, is, similar to the macaques, a signaling device. The hindquarter patches are, however, more pronounced than those of the macaques. The tail rises sharply from its base for some of its length and then droops.

These relatively big animals require large quantities of food, and so the often arduous search for edible fare occupies nearly the whole day and keeps them constantly on the move, rolling away stones and tearing out handfuls of grass. In arid zones they seem to be able to divine the presence of water, sometimes scraping away the surface soil to reach underground springs.

Baboons (*Papio* spec.)
grooming each other.

They are extremely fond of delicacies such as ostrich eggs which they dexterously crack open, sucking out the contents, or honey and the larvae of bees and wasps. They are also known to catch and consume young vervet monkeys, hares, newborn or slower moving gazelles. Leopards, lions, dogs, rock pythons, and crowned eagles are their enemies, although with their strength and mobility the baboons are well able to defend themselves, inflicting severe or even mortal wounds with their ferocious-looking canines on attackers, including leopards.

Gestation lasts about 190 days and only one young is born at a time. In the wild they rarely live longer than 25 years; an age of 37 years has been recorded in a few cases (Chicago Zoo). The baboons are distributed over vast stretches of Africa south of the Sahara with the exception of tropical forest zones. In addition, hamadryas baboons are found in the coastal regions of southeastern Arabia. Baboons are animals of the plains and savannahs. They adapt themselves easily; neither human settlements nor cultivation of the soil can oust them.

Their evolutionary adaptation to the far harder conditions of life on the ground in the open country comprised the acquisition of specific behavior patterns and a highly developed social order. But this higher evolutionary process led them further away from the ancestral simian type than the trend which ultimately produced human beings. The division, in the course of which the macaques and guenons also took the path of the baboons, happened at the beginning of the early Tertiary period about 25 million years ago. But within the continuous process of primate evolution the baboons attained a different kind of leading position.

Together with the macaques and hanuman monkeys wildlife baboons have been the subject of more exhaustive investigation than any other monkeys. During the nineteen-sixties two American anthropologists, S. L. Washburn and I. De Vore, made a systematic study of the social life of baboons in the grassland and savannahs and the structure of their societies in East Africa. They were convinced that knowledge about the behavior and habits of these monkeys would provide evidence about the origin and evolution of human society. At the beginning of the Pliocene period—about 12 million years ago—there was a change of climate in Africa too. It became notably drier; many forests gave way to grassland and savannahs so that some of the primates were faced with the need to adapt themselves to the new conditions, with all the dangers they involved in the struggle for existence. The two scientists observed more than 30 simian groups, mostly baboons. Their social system is organized in a similar way to that of the macaques. Association between male and female does not change during the time of their age-conditioned biological potency and is not dependent on the cycle of sexual receptivity; sexuality plays an important part in social relationships. Washburn and De Vore noted that social unity is one of the main factors determining the survival of the species, since a baboon spends nearly all its life in the close proximity of its fellow monkeys. It can only find security in the unity of the troop, and the baboon communities travel through their territory as a unit, whereby a herd of baboons is not merely a conglomeration of individual animals; it also constitutes a union of widely differing animal personalities. The size of these units differs from place to place; groups numbering from 12 to more than 200 animals were observed. Washburn and De Vore also noted that a troop of baboons may roam through an area of from 5 to 15 kilometers but only makes real use of specific parts of this territory. When water and food are widely scattered, the troops hardly ever meet. The territories of neighboring herds nevertheless overlap to a great extent. This could be best confirmed in the Amboseli National Park towards the end of the dry season. Water was only available in certain places; and various troops often came to the same water-hole to drink and eat the juicy plants in its vicinity. On one occasion we counted more than 400 baboons all gathered round one water-hole at the same time. A casual observer would have taken them for a single herd; actually three large troops were feeding side by side. They came and went, without mingling, although they fed or sat in close proximity. The scientists saw no battles between the groups at the water-holes.

Apart from their declared enemies, baboons pay little heed to other animals, nor do they usually take much notice of human beings. In open country the close association of various animal species up to the point of symbiosis is generally advantageous to all concerned. Hoofed animals, gepards, hyenas, and jackals have been observed within sight of baboons without any signs of nervousness or alarm leading to flight on the part of the latter. Quick reaction to a potential danger requires a very good group discipline. This is demonstrated, for instance, when a baboon troop descends from the shelter of the trees to the hazards of the open plains. The procession is headed by powerful males and stronger young males who act as the vanguard. Together with the adolescents, non-nursing and older females they also guard the flanks. The rear is brought up by a group of similar structure. The middle of the group is composed of the high-ranking, strongest males who are responsible for protecting the mothers and children. In this order they slowly move through the savannah in search of food.

Adult males express threatening behavior by a ferocious yawn exposing the throat and intentionally displaying the powerful ivory-colored canine teeth. The punishment of inferiors is adjusted to the severity of the offence. The mildest punishment is the so-called "black look." The high-ranking animal suddenly directs a wide-eyed stare at the culprit, who usually retreats to a safe distance with all the signs of panic. If this has no effect, the offender is seized by the scruff of the neck and whirled around in the dirt. The worst punishments are neck bites and expulsion from the clique. Usually, submissive gestures prevent serious injuries. The vanquished baboon surrenders by crouching down with his back to the victor in the symbolical posture of a sexually receptive female. Amongst the baboons, too, this pose inhibits the desire to attack. The ritual mounting, which, however, has no sexual significance, may take place as a gesture of self-assertiveness, but also as one of submission. Humiliated males beat a hasty retreat and, for the time being, avoid an encounter with their lord and master. Even when the

physical strength of high-ranking old males begins to diminish and they have perhaps lost their canine teeth in battle, they still often retain authority as a result of their ingrained self-confidence. When resting, small troops within the group devote themselves almost entirely to the ritual of grooming whereby the highest ranking males, identifiable by their thick manes, and mothers with young children, receive the most attention, often being surrounded by individuals anxious to perform this service. The high-ranking males usually act as wardens, settling conflicts, protecting the weak, taking care of the adolescents and children, hastening to give aid and meting out punishment by means of threatening gestures, cuffs and, if necessary, bites.

The unity of the society is not primarily dependent on sexuality, as was assumed up to the early nineteen-sixties. It is the instinctive and acquired pattern of social contact that determines their whole way of life. Each individual's need for contact is the major force, even in the case of low-status animals under the severe rule of dominant males. De Vore and Washburn noted that the baboons' social behavior does not follow instinctive patterns but is based on a surprising adaptability to specific situations. Sexually receptive females are more active and display more temperament, severing contact with the other females to a large extent, keeping close to the males, who also pay increased attention to them, until at the peak of their readiness to mate the dominant males claim them for copulation. The couples stay together only for a short term. No male stays with several sexually receptive females at the same time and keeps none in his sole possession for any length of time. The baboons therefore have no genuine and lasting family although many individual scenes in their social life may arouse this impression. As a rule the females never leave the groups into which they were born and reared; on the other hand young adult males sometimes do so. If a group grows too large in the course of time, this may cause a lack of stability. Supervision becomes impossible and a division is inevitable. This usually happens in a peaceful manner.

Like other animals living on the ground in open country, baboons fear the unknown. They can only overcome this constant anxiety through an exact knowledge of their territory and they start to acquire this at an early age. When darkness begins to fall they seek their sleeping-places under which enemies may lurk in the night-time. The large cushion-like hindquarter swellings make it easier for the baboons to sleep in a sitting position on the hard branches and rocks. Jostling for good places to sleep, a false step or loss of balance may spell disaster for an individual animal. Once they have fallen, the animals rarely find another safe sleeping-place that night. When daylight has returned the baboons very cautiously leave their elevated sleeping-places in smaller groups, having first reconnoitered the terrain.

The variability and adaptability of the baboons has enabled men to use their services, in a similar way to the pig-tailed macaques. Early civilizations, the Egyptians in particular, paid great attention to the baboons, even revering them as gods. In most zoological gardens baboons are usually found in large groups and in open-air enclosures (rocks). Their social behavior makes them popular with visitors. Nowadays the baboons are classified into five species. During the Diluvian period baboons lived in India and China too. The existing species are classified into two groups:

Baboons: Some zoologists classify them as one species with several subspecies. **Hamadryas Baboons:** Because of distinct deviations some zoologists classify them as a special subgenus. Only one species.

Chacma Baboon *(Papio ursinus)*. This is the largest baboon species; the body length exceeds one meter in some cases, whereby the female is much smaller, as is usual with monkeys living on the ground. The fur is grayish brown, the face, hands and feet are black. Corresponding to the geographical distribution, the coat varies in color to a greenish brown. The sparse beard leaves the ears free. The range encompasses South Africa to the north of Zimbabwe, Namibia and southern Mozambique. They preferably live in dry, rocky regions. Inaccessible cliffs which guarantee safety at night are their favorite sleeping-places. Like all baboons, they can be trained to work for human beings. According to accounts by Hoesch, young female chacma baboons were used as goatherds after a brief period of training. "After only a few days these 'apprentices' accompanied the goats when they went out to graze; they had adapted themselves to the alien herd companions. They kept the goat herds together, preventing them from straying and in the evening took care that they returned together to the kraal. The farmer (Mrs. Aston) did not have to train them to do this; the baboons simply followed their sociable disposition. One of the female baboons was even able to bring the lambs born on the pasture land back to the farm in the evening. Young baboons normally cling to their mother's fur and are thus carried instinctively by her. A baboon mother only carries her baby under her arm—in order not to lose it—when in flight or in dangerous situations. But here it was a seemingly logical reaction on the part of the baboon. Without previous experience she obeyed her maternal instincts in adopting the only practical course of carrying the lambs in this alien manner under her arm.

Chacma baboons are not common in zoos.

Four subspecies, which vary respectively in the color of their fur, have been identified.

Yellow Baboon *(Papio cynocephalus)*. More slender and long-limbed in its proportions, it varies greatly in type and in the color of its coat which ranges from grayish yellow, light brown to olive-green; the face is a blackish lead gray to a dark flesh color; body length up to 75 cm; shoulder height 50 or 60 cm. The females remain smaller. Its range adjoins that of the chacma baboon in the north and extends approximately from the Zambesi to the southern fringe of the forest-clad Congo Basin as well as from Uganda and Kenya to the south of Somalia. It lives predominantly in the savannah, in scrubland, but also in clearings and open wooded country. The herds often do considerable damage to plantations.

The yellow baboon is uncommon in zoos although breeding successes have been achieved with smaller groups.

Five subspecies have been identified which also display considerable variations, for instance, *Papio cynocephalus ruhei* in Somalia is gray green while *Papio cynocephalus ochraceus* is more brownish in color.

Anubis or **Olive Baboon** *(Papio anubis)*. The name is derived from the dog or jackal-headed Death God of the Ancient Egyptians. This baboon's body is more heavily built. The males weigh about 33 kilos, the females only from 17 to 20 kilos. The male's body length lies between 80 and 95 cm, the tail is about 45 cm long. Gestation is said to last 183 days. The adult male is furnished with a thick mane on its nape and shoulders. The fur varies from gray to olive-green with finely spotted markings. The face is dark gray to brownish flesh-colored. The anubis baboons inhabit a zone north of the large forest belt extending from the Niger to the Nile towards Ethiopia and Somalia. They are found very locally in the Sahara, for instance in the region of Tibesti in northern Chad, and in the northwestern part of Sudan. Their environment consists of savannahs and scrubland where they often seek their sustenance among herds of zebra and antelopes such as Grant gazelles and impalas. When a favorable opportunity occurs they steal newborn gazelles and eat them.

Anubis baboons are common in zoos.

The number of subspecies is variously specified; usually seven are named.

Guinea Baboon *(Papio papio)*. This is the smallest species, the male rarely exceeding 65 cm in body length and 28 kilos in weight. With slight variations the fur is light to reddish brown in color. Nape, shoulders and back are covered by longer fur. The face is dark to light gray. This baboon inhabits a more restricted savannah and grassland zone in West Africa extending from Senegal and Guinea via southern Mali, northern Upper Volta to the northern part of Nigeria. In zoological gardens it is not very common, but it has been bred in captivity. There are two subspecies.

Hamadryas or **Sacred Baboon** *(Papio hamadryas)*. The most conspicuous and astounding sight among all the true baboons is the fully adult male hamadryas. Its body length may reach 76 cm with a shoulder height of up to 55 cm. Old males can even exceed these measurements. Their weight is rarely in excess of 30 kilos. The grayish brown females are smaller, and generally weigh under 30 kilos. The face has a well-developed muzzle with ferocious teeth, is naked and grayish yellow to flesh-colored. The large cushions of flesh on the hindquarters are bright scarlet and, when the animal is well-fed, filled with fatty tissue. When the female is sexually receptive she has large red swellings under the tail that look like malignant tumors but are, in fact, quite normal.

In Ancient Egypt the hamadryas was worshipped as a sacred animal and was represented on numerous murals and as sculptures. It enjoyed great respect and privileges as the companion and servant of the moon god Thoth, being regarded as a priest directly descended from the gods. When they died these baboons were mummified and buried in special tombs. Like the pig-tailed macaques today, they were trained to pick fruit. It is not certain whether in the days of the Pharaohs the hamadryas baboons inhabited parts of Egypt or were brought from the Sudan by river to the lower valley of the Nile.

The rock-dwelling hamadryas baboons have adapted themselves successfully to their austere environment by evolving group behavior that differs from that of the baboons living in the plains. In contrast to all other baboon species their groups are divided up into definite sub-groups and these again into smaller troops. The Swiss zoologists H. Kummer and F. Kurt (1965, 1968) observed hamadryas baboons for several years and came to some instructive conclusions. There are four distinct units: the largest is the herd usually consisting of several hundred animals; it is divided up into the troop with about 50 to 100 animals, the clan with ca. 20 to 30 and the one-male group with three to nine baboons. Apart from their common sleeping-places on the rocks and a larger territory through which they jointly roam during the daytime, nothing else holds the herd together. Suitable rocks that also guarantee safety from leopards are few and far between, so that sometimes a herd numbering 700 to 800 and more animals is crowded together on inaccessible rocky ground. The one-male group is composed of two or more females and young animals and is led by an adult male. Since the females can reproduce at an earlier age than the males and their mortality rate is lower, there are about two adult females to every male. The females of older group leaders are taken over by younger males—usually not without a battle. There are no free or independent females; they all belong to a one-male group. Each group leader almost jealously safeguards the integrity of his community. If a female strays too far away or attempts to transfer to another group, the leader tries to dissuade the truant by threatening behavior, to begin with. If she disregards this unmistakable warning, the angry leader bites her in the scruff of her neck. This special kind of shep-

herding is only practised by the hamadryas baboons, which, however, restrict it to members of their own group. Strange females in another group are taboo for a male, even if he has no mate. The one-male groups hence possess the closest and firmest social bonds. The association between male and female is lasting and not dependent on the sexually receptive cycle; sexuality itself is not limited to mating but is of great importance in terms of social relationships. In a ritualized form this applies to submissive, welcoming and contact behavior. The following is also typical for the hamadryas baboons: a lower-ranking male adopts a submissive attitude towards a dominant male while simultaneously threatening a third baboon. In this manner the second male enjoys protection and directs the attention and resulting aggressiveness of the dominant male towards the third baboon who is perhaps his rival in the status struggle.

The range of the hamadryas baboons is nowadays limited to the uplands in the northern and eastern parts of Ethiopia and Somalia as well as a narrow mountainous coastal strip in southern Arabia—zones with sparse bushes and only scattered tree cover, with wadis and rocky ground in parts.

Their main fare—buds, blossoms, leaves, grass, roots, tubers, and fruit—is scanty and scattered in small quantities over the meager vegetation which is consequently intensively utilized by only small foraging groups. Ponds and rocks to sleep on which, although limited in number, are large enough to provide water and protection for many animals, call for the ability to form both small and very much larger groups as circumstances demand. The dominating conditions here demand extreme adaptability which is naturally dependent on genetic endowments.

The hamadryas is the most common baboon in zoological gardens and is sometimes also exhibited in smaller enclosures where, since it avoids water as far as possible, a shallow canal provides an adequate barrier. Their maximum age is about 34 years.

The only recorded subspecies is the somewhat smaller hamadryas baboon that lives in southern Arabia.

Mandrills (genus *Mandrillus*)

Living like the true baboons mostly on the ground, these animals are, however, forest-dwellers. Common to the species of this genus is a vertically elongated anterior skull with distinct bony ridges covered with furrowed skin on each side of the muzzle. The head of the powerful and muscular animal is relatively large and gives an impression of heaviness; the muzzle is massive with large forward-facing nostrils; the eyes are close-set. Some parts of the skin are highly colored. The tail, which is reduced to a stump, is carried upright. Primatologists regard the species of the genus *Mandrillus* as representatives of an evolutionary trend amongst the baboons in the course of which, due to the loss of tree cover when the climate changed, they adapted themselves physically to life on the ground in open country. Later on, they again retreated to the forests, but retained the habits of an animal living on the ground; they were well equipped for this kind of life by their physical strength and ability to defend themselves.

Mandrill *(Mandrillus sphinx).* The body length of the male can measure up to 90 cm, the shoulder height even exceeding 50 cm, with a weight of up to 50 kilos; the tail is 6 to 9 cm long. The face coloring is probably the most spectacular among the mammals: the bridge and front part of the nose are scarlet while the furrowed parts of the cheeks are bright cobalt blue. The sparse moustache as well as the more thickly bewhiskered chin and cheeks are whitish yellow to orange. On the side of the head there is a collar of the same color. Parts of the forehead, shoulders, chest and neck are of a deeper hue. There is a white patch behind the pointed ears; the fur is longer on the top of the head and neck. The rest of the fur is dark grayish brown with greenish tints on the upper side and whitish to gray-brown on the underside. The skin round the anal region is of an equally startling coloring: deep red and bright blue with violet tinges. The hindquarter patches and scrotum are also red. So far zoologists have found no completely satisfactory explanation for this strangely colored signaling device. When challenging a potential attacker—for instance a leopard—the male mandrill, who usually maintains a distance from the group but keeps within sight of it, stands upright ready to pounce, bending and stretching his muscular raised arms, staring unswervingly at the head of his enemy, and threatening by opening his mouth to its fullest extent and displaying the long, dagger-like canine teeth. In this highly agitated state the colors of his body deepen in hue, an important clue to their significance. Seen from the front, the chest, too, is bright blue, and red and blue patches appear on the arms, hands and feet. These glaring colors visually enhance the ferocious impact of the imminent onslaught and possibly urge the enemy to seek safety in flight. In this connection it is interesting to note that among these monkeys even deep wounds or extensive surface injuries rarely become infected. The colored skin on the hindquarters has a different significance. The Swiss naturalist Konrad Gesner noted: "If you threaten him with your finger, he presents his backside to you." This demonstrative gesture of submission—the presentation of the anal region—is common to many Old World monkeys. Originally the pose of a sexually receptive female, in the course of the evolutionary process this attitude attained a different significance, and in its ritualized form serves to placate higher-ranking animals. The garish colors are also interpreted as a signaling device with which the adult mandrill in the dim light of the forest optically proclaims his presence to males belonging to other groups. Nor does the display of an erected penis proclaim sexual intentions but may be interpreted as the optical marking of the group's territory (W. Wickler 1967) or as an indication of the presence of authority on the part of dominant males.

Relatively little is known about the mandrill's life in the wild. Its comparatively small range is situated in West Africa, mainly in Cameroun, and in narrow strips of the neighboring states. It inhabits rain forests where the undergrowth is less dense, seeking mostly vegetarian fare on the ground between the trees and also small animals which it finds under stones and fallen tree trunks. When danger approaches it usually seeks refuge on the forest floor; it only gives preference to safe places on the trees during the night-time.

Mandrills have a group size and social structure similar to that of their cousins in the savannah. The smallest unit is normally composed of an adult male as group leader and nine to twelve females, younger males and offspring. They always maintain vocal contact with each other. The meaning of the different kinds of gestures and play of features, particularly of the male mandrill, has been the subject of many studies: the playful displaying of teeth by raising the corners of its lips, chattering teeth, the nodding or shaking of the head as a sign of friendly contact or welcome, or beating the ground with the hand as a sign of irritation; the smacking of lips, threatening yawns, bristling hair, the display of muscular power, etc. Its powers of self-defence and boldness allow the mandrill to visit smaller clearings and the fringe of the forest too. A leopard—the mandrill's main enemy—that survived an encounter with this animal would heed the deterrent signals and would be unlikely to risk fresh attacks. The mandrills avoid snakes and only young animals that have strayed into forest clearings are preyed upon by the crowned eagle.

Mandrills are fairly common in zoological gardens where their startling coloring makes them a major attraction. Isolated animals or those in groups that are not sufficiently accustomed to each other display a marked tendency to disturbed behavior. Because the mandrill is often hunted there is a serious depletion of the herds.

Drill *(Mandrillus leucophaeus).* It resembles the mandrill in many ways. The drill is somewhat smaller and more slender, its body measuring about 80 cm. The head looks very powerful. There is a bony ridge which is, however, not covered by furrowed skin, on each side of the nose. The jutting cheek-bones widen the shiny black face. Only the chin is a conspicuous red. The head is surrounded by a mane of long, fine, off-white fur which is darker at the tips; the chin and whiskers are of the same color; the hair on its forehead is darker. Hindquarter patches and scrotum are scarlet. The body is gray to brown with olive-green tints; the underside is white to gray. Its range largely coincides with that of its nearest relative but extends south-wards to the mouth of the Congo. The habits and behavior of the two species are also very similar. The drill is less confident in its manner than the mandrill; it appears shyer and more reserved. According to T.T.Struhsaker a drill group may number from nine to 55 animals and usually averages about 24. In captivity drills thrive just as well as the plains-living baboons, but are less common in zoos than mandrills.

Geladas (genus *Theropithecus*)

Gelada Baboon (*Theropithecus gelada*). Although, generally speaking, the gelada greatly resembles the baboons, it is classi-fied as a separate genus. Its body measures about 70 cm, the tail with tuft ca. 55 cm, and its shoulder height is up to 55 cm. Its weight ranges between 25 and 28 kilos; the much smaller fe-males only weigh half as much. The body appears slender and comparatively long-limbed. The head is shorter and the ante-rior skull, similar to that of the drill, more rounded than in the case of the baboons. A bony ridge runs along the cheek-bones from the muzzle to the outer corner of the eyes. The light to medium brown beard that stands out sideways extends almost to the forehead, making the head appear larger than it is. The front part of the body is furnished with a dark brown, long-haired cloak-like mane. The female, whose fur is also brown, has a conspicuous long pointed crest.

The geladas are only distributed over a small area on the partly rocky high plateau of Ethiopia extending from the north of Addis Ababa nearly as far as the Red Sea. In the north and east it partially overlaps the range of the hamadryas baboons. As mountain-dwellers, living locally at an altitude of from 3,000 to 4,000 meters, the geladas have adapted themselves to the harsh environment. Their thick coat protects them from the cold at night. Their very strong curved finger-nails enable them to extract food from the hard ground; they also eat grass as well as leaves and naturally all kinds of small animals. Before dusk they seek safe sleeping-places on sheltered cliffs with a sheer drop to deep valleys.

According to observations made by J.H.Crook and P. Ald-rich-Blake (1968), the one-male group is the basic social unit and consists of the dominant male, several females and young animals—a total of between nine and ten members. Sometimes exclusively male groups are found. Away from their sleeping-places on the rocks, groups numbering up to 400 individuals have been observed. This is more often the case during the rainy season when, as a result of abundant vegetation, food is plentiful. In the dry season, the small one-male group offers better opportunities for subsisting on the scanty food supply. The geladas cover a distance of up to 7 kilometers daily. They quench their thirst from time to time with water found in rock

crevices. Much care is lavished on the children, by the males too. Rivalry between high-ranking males rarely leads to severe bites and is generally limited to threatening behavior which, however, has a terrifying effect. W.Fiedler (1967) describes it in the following way: "When threatening, the geladas fold back their upper lip so that the bright inner side can be seen, thus baring their teeth with the huge canines. At the same time they raise their eyebrows, revealing their gleaming white eyelids which stand out in stark contrast to the black face. The naked reddish skin on the chest turns scarlet when the animal is great-ly agitated." At such times the gelada makes a truly ferocious impression. The male's red anal region to which the small grayish black hindquarter patches form a contrast, resembles in shape and color the female's anogenital zone. When the fe-male is sexually receptive this zone inflates into bright scarlet puffy swellings. The males, however, permanently display this conspicuous sign, most probably as a demonstration of social status. It possibly also carries significance for the mounting of submissive low-ranking males. The unusual red patch on the chest deepens in color and is also surrounded by a fairly distinct chain of white blisters when the animal is sexually receptive. Afterwards this signal fades. Mating and the ensuing births are restricted to a seasonal cycle. The gelada is not very common in zoological gardens since it requires conditions approximating to its mountainous habitat. The black gelada (*Theropithecus obscurus*) classified by H.-W. Smolik (1960) is probably a color variety.

Mangabeys (genus *Cercocebus*)

The species of the following two genera of long-tailed monkeys are, with one exception, definite tree-dwellers. As a result of adaptation to this way of life they possess more slender hands and feet, longer tails and also the ability to clamber nimbly and swiftly and to leap well. The body length ranges from 40 to 70 cm, the tail is from 45 to 75 cm long, the shoulder height is about 48 cm and they weigh from 5 to 15 kilos. They are akin to the macaques but have a type of body and habits that resemble those of large guenons. The anterior skull is longer, the eye-brow ridges are ledge-like. They have powerful jaws with large canines and five protuberances on the lower back molars. Hastily eaten food is stored in the large cheek pouches. The voice is not particularly loud because the larynx resonator is not highly de-veloped. The long tail is carried slanting upward from the base and also sometimes pointed forward over the back. It has a balancing function while clambering, jumping and running along the branches and is also thought to play a certain part in regulating the body temperature in very hot weather. Under the tail and on the buttocks there are naked hard swellings, which in the case of older males may grow larger, but by no means reach the dimensions of those on the baboons' hindquarters. The sexually receptive female's genitals are only slightly swollen and are less highly colored than those of the baboons. These monkeys are omnivorous but fruit is their main source of nourishment; they also eat green shoots, buds, and —on the rare occasions when they come down to the ground— tubers and seeds. They hunt insects, tree frogs and small lizards too, and eat birds' eggs with relish.

The mangabeys mainly inhabit the wet tropical forests of Central and West Africa. Although not abounding in numbers, the individual species are widely distributed, particularly in the river catchment zones. They sleep in the middle and lower tiers of the tree cover.

Their social bonds are so far relatively unknown because in the dense forest with the confusing play of light and shade on the foliage individual animals are very hard to spot and their distinctive features can only be identified in the course of lengthy and systematic observations. They are also shyer, quieter and more reserved than the loud guenons. An interesting feature is the way they communicate; the calls uttered by the few species with loud voices are coupled with eloquent facial expressions.

The genus is subdivided into four distinct species each of which has several subspecies.

Sooty Mangabey *(Cercocebus torquatus)*. With a body length of up to 70 cm and a weight of up to 15 kilos this is the largest mangabey species. The color and marking of the fur differ from subspecies to subspecies. The face is dark gray to black, forehead and top of the head are olive to red-brown; back, flanks and limbs are gray to brown on the outside, abdomen and insides of the legs and arms as well as the tip of the tail are light gray to white; brows, neck and nape are white. Their range in West Africa extends from Sierra Leone to Ghana and there is also a remote region from the mouth of the Niger to that of the Congo, where they are chiefly found in swampy forests, wet gallery forests and even in mangrove swamps and near river mouths. They often come down to the ground, are not fussy about what they eat and even encroach on cultivated land, sometimes causing damage to ricefields and other plantations. In Sierra Leone they are regarded as most harmful and are said to endanger the cocoa harvest. Such biassed views should, of course, be treated with caution since they may easily unleash an extermination campaign. Mutual grooming plays an important part in these monkeys' social behavior. They are relatively uncommon in zoos. Three subspecies have been classified.

Gray-cheeked Mangabey *(Cercocebus albigena)*. This somewhat smaller species has a shaggy long-haired coat that varies on the upper side in particular from smoky to dark gray without any conspicuous markings; nape, cheeks, and chest are somewhat lighter in color. In contrast to the other species these animals have almost fully developed larynx resonators which enable especially the males to utter loud rumbling calls that are accompanied by grimaces, a bristling crest and a bent back with a vertically held tail. Their range extends from West to Central Africa. They are uncommon in zoos. The three subspecies are:

Cercocebus albigena albigena, southern Cameroun, Gabon and Lower Congo; *Cercocebus albigena zenkeri*, named after the German explorer Zenker. Central Cameroun; *Cercocebus albigena johnstoni*, eastern Cameroun to the Ituri river and Lake Mweru on the border between Zaire and Zambia.

Black Mangabey *(Cercocebus aterrimus)*. Can be identified by the long, pointed upright crest on the top of its head. The black, long-haired shaggy coat has no markings and only the whiskers are of a lighter grayish brown shade. Other features are a jutting face and light eyelids. The hands and feet appear relatively large. This species comes down to the ground more frequently and seeks sustenance near the banks of rivers, but only where there is adequate tree cover. In some places they share their range with sooty and gray-cheeked mangabeys. They inhabit the belt of tropical forest within the great bend of the Congo in Zaire that extends to the northwestern tip of Angola. This species is seldom found in zoological gardens.

Agile Mangabey *(Cercocebus galeritus)*. This species is of medium size. Coloring and markings of the subspecies vary a good deal. The basic color is brown to olive-green, the fur being speckled with yellow or golden brown tints, while the chest and underside are lighter in color. The fur on the head is longer, forming a flat cap that may be adorned with a curly tuft. The whiskers are shorter than those of the other species. Their range encompasses Cameroun, except for the northern part, the northeast of the People's Republic of Congo, the Central African Republic as well as Zaire with the exception of the southern part of that country. They are also found to a much lesser degree in East Africa, including the forests flanking the middle and lower reaches of the river Tana (Kenya). Part of this area in which colobus monkeys also live, has been given the status of a reserve. This species is occasionally found in zoos where they usually breed regularly and have a good survival rate. The following subspecies are named here:

Cercocebus galeritus galeritus, head tuft and curly fur on the forehead. Tana river zone; the easternmost form; *Cercocebus galeritus agilis*, body and head fur heavily speckled with yellow; crested tuft. South Cameroun via Zaire north of the Congo bend to the river Ituri in the east; *Cercocebus galeritus chrysogaster*, fur on the head falls toward the back. Central and southern parts of Zaire; *Cercocebus galeritus hagenbecki*, a very light-colored variation that is only classified by some authors as a distinct form.

Guenons (genus Cercopithecus)

True guenons are small to medium-size slenderly built monkeys with long tails (cercus = tail, pithecus = monkey). The body measures from 30 to 70 cm; the tail is from 50 to 85 cm long. Some species weigh slightly more than 10 kilos. They have a life expectancy of up to 26 years. The guenons only have small hindquarter patches. They can store large amounts of food in their cheek pouches. Furnished with strong dagger-like canine teeth, the males in particular have an intimidating appearance.

Guenons are conspicuous for their surprisingly large variety of colors and fur markings, which makes it difficult to classify them, especially since there are sometimes transitions without any clear limitation. The naked or thinly haired parts of the bodies are also very brightly colored; the males of some species have garishly colored scrotums. Often the fur coloring of the young animal is different from that of the adult—a fact which formerly sometimes gave rise to the erroneous identification of new subspecies or even species.

Nowadays the guenons are classified into 14 species; according to Sanderson the subspecies so far identified amount to 72. The somewhat different patas may be added to this number. The African guenons therefore form one of the largest primate groups with approximately 90 varieties. As true tree-dwellers they are distributed over almost the entire continent south of the Sahara. Each region has its own guenon population; many live in belts of tropical forest both in the lowlands and uplands. In addition certain species are found in specific types of tree cover as well as in specific tiers of the tropical forest. In the open country, savannahs and gallery forests in tropical grasslands are favorite habitats. "When there are trees there are also guenons" (Sanderson).

The most active phases of their daily lives are the mornings and afternoons. During the midday heat they rest, spending much of their time in mutual grooming. The guenons' need for social contact is not so pronounced as that of the macaques and baboons. But here too there are, of course, specific gestures signifying welcome, threats, display, fear, and submissive behavior. The guenons have no clearly defined order of precedence or social system such as is common to most primates living on the ground in the plains. Nevertheless, older and more experienced males do carry out certain leadership and guardianship functions especially when the herd temporarily leaves its usual territory.

In loose-knit groups ranging in size from 25 to far more than 100 animals the lively and restless guenons roam their territory on the alert for food. Large marauding groups that frequently plunder areas under cultivation can become a regular plague particularly in view of the fact that the guenons, like all monkeys, pluck more fruit than they need to satisfy their appetite, discarding it after a few bites.

They maintain contact with a large repertoire of calls. When disturbed, they remain quiet and motionless to begin with, often pressing their bodies against the branches and tree trunks. Their eyesight and hearing are well developed. If one of the animals scents danger it at once utters a loud barking sound in which the others excitedly join. When the danger has been located they close their ranks and scamper away in the opposite direction. The long and relatively heavy tail has an aerodynamic function when jumping and acts as a "balancing pole" amongst the branches.

Their territorial dependence varies, being influenced by several factors. Whereas certain species or groups stay in a specific territory, others roam through a larger region or even leave their original territory. Another species can then fill the gap left by their departure. So the areas where some of the guenon species or populations occur shift within the limits of their habitat. Moreover, growing economic encroachments on the wilderness are causing further substantial changes. Some of the guenon species, however, are adapting themselves to this new environment, no longer avoiding the presence of human beings —provided they suffer no definite harm from them—and more or less exploiting the plantations as an abundant source of food.

Only a few guenon species would appear to have a definite reproductive cycle. Young are born all the year round after a gestation period lasting from 6.5 to 7 months. The mothers carry their young ventrally. To begin with, the baby obviously cannot cling firmly or safely enough to the mother's fur, so she holds it close to her with her hand, especially when on the move. Later on the baby no longer requires this additional support. Development and growth progress rapidly during the first months of life. Under the watchful eye of the mother the young begin to take solids at the age of from three to five weeks. The guenons start to reproduce at the age of about 4.5 to 5 years. Their natural enemies are leopards, African wild cats, pythons and the larger birds of prey.

The lively and active guenons are very popular animals in zoological gardens and their bright coloring and conspicuous markings attract much attention. When properly cared for in captivity they thrive and have a good life expectancy. Hand-reared young animals can become very tame, but only attach themselves to those people who understand their individual ways. Isolated females remain relatively tame even when adult, although guenons usually become less amicable with adulthood and are often inclined to snap.

Breeding successes are uncommon as a rule. There are several instances of cross-breeding between various guenon species or subspecies. When kept in confined quarters older males become quarrelsome and only tolerate one or two chosen adult females as their partners. Guenons are not suitable domestic pets.

Vervet or **Grass Monkey** *(Cercopithecus aethiops)*. The best-known and probably most numerous species of this genus. Actually these monkeys are all yellowish brown, some with light grayish shades and more or less deep olive-green tints that encompass a broad spectrum of variations. The abdomen and inner surface of the limbs are lighter or whitish. Equally light to whitish broad forehead stripes and beards are further features. The face is black. The body measures from 40 to 58 cm, the tail is from 50 to 70 cm long and they weigh between 5 and 7 kilos. Various sources state that their life expectancy does not exceed 25 years. All vervet monkeys have legs that are longer than their arms. On the ground they move in a similar fashion to dogs. In high grass the tail is carried vertically, no doubt as an optical signal for mutual recognition and keeping in touch with the herd. If they run along the branches the tail is held horizontally, the tip drooping downward; the tail apparently has a balancing function.

Vervet monkeys inhabit the tree-covered grassland south of the Sahara, from Somalia to the Cape in the south and as far as Angola. Hence they are the most widely distributed guenon species. This vast range is not cohesive but broken up into various larger and smaller zones. The prominent mammal specialists D. Starck, H. Frick and W. Fiedler speak of a "characteristic animal of the bush, in particular the fringe of the rain forests, the herds are always found in proximity to waterways; they avoid open dry grassland but also the interior of forest belts . . . Where they find suitable living conditions they usually live in herds numbering from 20 to 50 animals; we occasionally also saw smaller troops and individual animals which possibly had only temporarily left their herd; we observed that these guenons preferably inhabit gallery forests flanking the course of rivers where each herd possesses its 'territory.'" Herds that roam extensively generally return to their original territory even after months have elapsed. Vervet monkeys can no longer be regarded as true tree-dwellers. They are equally nimble and mobile on the ground and on rocks. As a result of the previously mentioned wide distribution and the varying ecological conditions, about 20 geographic subspecies have evolved. If they have suffered no harm at the hands of human beings or are used to their proximity—for instance on the fringe of settlements and in reservations—they approach to a distance of 10 or even 5 meters.

As an omnivore the vervet monkey's diet consists to about fifty percent of small vertebrates, insects, spiders and other animal prey. Sometimes it takes eggs from birds' nests. Its vegetarian diet is made up of fruit, onions, various kinds of seed, tubers, roots, young leaves, and buds. The food is often hastily stored in the cheek pouches and then chewed and eaten at leisure. These guenons have no lasting consort relationships. Both sexes are usually polygamous. There is little serious rivalry amongst the males.

According to observations made by D. Chaney and R. Seyfarth (1980) in the Amboseli Park in Kenya the individual sounds uttered by vervet monkeys contain more exact information than was hitherto assumed. There are warning sounds that differ according to whether danger takes the form of a beast of prey on the ground, a snake or a bird and even distinguishing between individual species. Young animals sometimes sound the leopard alarm on the appearance of lions or hyenas, and herons or storks are mistakenly registered as birds of prey.

Although in recent years very many vervet monkeys have been exported for experimental purposes, their numbers are not regarded as endangered. Nonetheless, this depletion of animals in the wild calls for restrictions and strict government control. Apart from regulations applying to reserves and national parks there are no effective protective measures.

The vervet monkey used to be the most common species in zoological gardens; nowadays more colorful varieties predominate. In the 18th and 19th centuries this species, along with rhesus monkeys, was a popular attraction in menageries where they were considered to be the embodiment of mischievous and droll simian antics. Of the many subspecies only the following are named here:

Cercopithecus aethiops aethiops, southern Central Africa, Sudan, Ethiopia; *Cercopithecus aethiops sabaeus*, Senegal to Ghana; *Cercopithecus aethiops cynosurus*, Zaire, southern Congo Basin, Angola; *Cercopithecus aethiops pygerythus*, East to South Africa; *Cercopithecus aethiops tantalus*, Ghana to Cameroun, Central Africa; *Cercopithecus aethiops callitrichus*, Senegambia to northern Liberia; introduced to the Cape Verde and Canary Islands and the West Indies.

Mona Monkey *(Cercopithecus mona)*. Mona guenons are true tree-dwellers and give preference to the tree-tops. "Mona" is the Moorish name for long-tailed monkeys. The main features of this group are: a dark blueish face with flesh-colored mouthparts, a white to yellow forehead stripe, beard of the same color, brown body with more or less greenish tints. The chest and inner surface of the otherwise dark limbs are light to whitish gray in color; athletic build. Body length: 45 to 55 cm, tail length: up to 75 cm, weight: 4 to 6 kilos. One of these guenons lived for 26 years in an American zoo. When living in the wild they probably die a natural death before reaching such an age. The many varieties are classified into seven subspecies which occur from Senegambia to the Congo Basin.

Although these monkeys do not seek deeper water of their own accord they are able swimmers if the need arises. Sanderson observed how a group tried to jump over a deep stream, but apparently misjudging the distance, fell one after the other into the water. They all cried out but paddled like dogs adroitly and swiftly on land, climbing up the trees there. Their diet is varied, consisting of about equal quantities of vegetarian and carnivorous fare. Their numbers are considered to be still stable. In zoological gardens with suitable tropical houses varieties of this species are fairly common. The better-known subspecies are:

Cercopithecus mona mona, Nigeria, northern Cameroun; *Cercopithecus mona campbelli*, Sierra Leone, Liberia; *Cercopithecus mona lowei*.

Crowned Guenon *(Cercopithecus pogonias).* This species is closely related to the mona guenons. They were previously classified as belonging to the mona forms but for some time now certain taxonomists have declared them to be a distinct species with four subspecies. These all closely resemble each other, having as a common feature a comb-like crest of fur down the middle of the head. The face is black, the mouth-parts flesh-colored. Two black stripes, beginning at the outer corner of each eye run across the brow, the fringes covering the ears. There is another similar stripe over the brows. The sides of the head are adorned by an upstanding beard of a sulphor or golden color. The back and sides of the body are a brownish hue with distinct leaf-green tints on the dorsal side. The origin of this green color, which is otherwise only seen on birds, is so far unknown. A dark wavy stripe runs along the back and tail. The chest and underside vary from deep yellow to orange. Hands and feet are black. So this guenon may be numbered amongst the most handsomely colored monkeys. Body measurements and weight are similar to those of the mona guenons. Their repertoire of sounds includes bird-like twittering calls.

The crowned guenon occurs from northern Cameroun—almost entirely restricted to mountainous regions—to the Congo area where it mainly inhabits the top parts of high trees. Some of the subspecies are only found on the fringe of forest belts, in gallery forests or in isolated larger stretches of tree cover in the savannah *(Cercopithecus pogonias nigripes).*

Sanderson found stomachs full of insects in a number of these guenons that had been recently killed so it may be assumed that, at least in some places, a large proportion of their diet is of animal origin. These guenons are not often found in zoological gardens. The subspecies include the following:

Cercopithecus pogonias pogonias, northern Cameroun and on Fernando Póo; *Cercopithecus pogonias grayi*, Cameroun, Equatorial Africa to the Congo; *Cercopithecus pogonias nigripes*, Gabon.

Diademed Guenon *(Cercopithecus mitis).* This guenon and its subspecies have various kinds of diadem-like patches of fur on the forehead. The body length of the males is from 50 to 65 cm, the tail is from 65 to 75 cm long and they weigh between 4 and 7 kilos. The females are definitely smaller. The body color is dark grayish brown to blueish gray with partly reddish brown or green tints on the back. The species can be identified by its light forehead or eyebrow stripes, white ruff on the chin and neck

and more or less long dark beards. One subspecies has a lighter head and a white throat.

These guenons inhabit an extensive region in Central, East and Southeast Africa as fas as the Indian Ocean. The true diademed guenons are only found in the interior of the continent. So only a minority of the species live in belts of tropical forest, the majority being found in savannahs with denser tree cover, gallery forests, wooded patches of tropical grassland, bamboo thickets and even in the more arid scrubland. As a result of this extensive range of ecological conditions about 20 subspecies have evolved, and a unified classification has not yet been arrived at. The habitat in which they exist compels these guenons to come down to the ground more frequently than their relatives in the depths of the virgin forest in order to find sufficient fruit, roots, tubers, seeds as well as insects and other small animals for subsistence.

The diademed monkeys are quiet and retiring by nature. They live in large social units that sometimes number more than 100 animals, roaming together in the course of one day through their extensive territory. More alert to danger than other species, they take advantage of every bit of natural cover to protect themselves. They are fairly uncommon in zoological gardens. Of the numerous subspecies the following are named:

Cercopithecus mitis mitis, northern Angola and south of the Congo; *Cercopithecus mitis doggetti*, Uganda; *Cercopithecus mitis albogularis*, East and Southeast Africa; *Cercopithecus mitis stuhlmanni*, East Africa; *Cercopithecus mitis kibonotenensis*, East Africa, region near Mt. Kilimanjaro and Meru.

L'Hoest's Monkey *(Cercopithecus l'hoesti).* The dense white or whitish gray beard forms a distinctive contrast to the dark gray to black coat which is also thickly haired. There are reddish brown or lighter saddle-like markings on the dorsal side. In some cases the white of the beard extends over the chin and throat to the lower part of the chest. With a body length ranging from 50 to 70 cm the tail is from 55 to 80 cm long. A noteworthy feature is the widely scattered and very local distribution of these guenons. They are the inhabitants of higher mountainous regions from the uplands of northern Cameroun over Ubangi Chari, the volcanoes at Lake Kiwu, eastern Zaire and Uganda to the slopes of Mt. Kilimanjaro and are typical forest-dwellers.

---

120 Hanuman or **Entellus Monkey**
(Presbytis entellus).
Hanuman monkeys alternate between wooded areas and human settlements and are often found in close proximity to villages and in suburbs. Human beings thus influence their habits to a considerable extent and help to protect the species. (Carl Hagenbeck Zoo)

121 Hanuman or **Entellus Monkey**
(*Presbytis entellus*).
Hanuman monkeys have various types of "hair-do"
which often take the form of upstanding tufts.
(Hanover Zoological Gardens)

122/123 *Presbytis entellus priam,*
a subspecies of the common hanuman monkey,
has a crest as its distinguishing feature.
(Hanover Zoological Gardens)

124/125 John's Langurs *(Presbytis johni)*.
As compared with the hanuman monkeys, far less
is known about the habits of John's langurs
in the wild. Their existence, however, is severely
threatened. (Hanover Zoological Gardens)

Overleaf:

126 John's Langur *(Presbytis johni)*.
Although the Hindus venerate this species too,
it does not occupy such a prominent position
as the hanuman monkey. Consequently these
langurs occur far less in the immediate
vicinity of human settlements and do not come
down to the ground very often.
(Hanover Zoological Gardens)

127 Dusky Leaf Monkey *(Presbytis obscurus)*.
Development in childhood and adolescence is
similar to that of the hanuman monkeys.
There is a lack of knowledge about the dusky
monkey's habits in the wild.
(Duisburg Zoological Gardens)

134/135 Banded Leaf Monkeys
(Presbytis melalophus).
In recent years several zoos have been more successful in their efforts with banded leaf monkeys.
(Cologne Zoological Gardens)

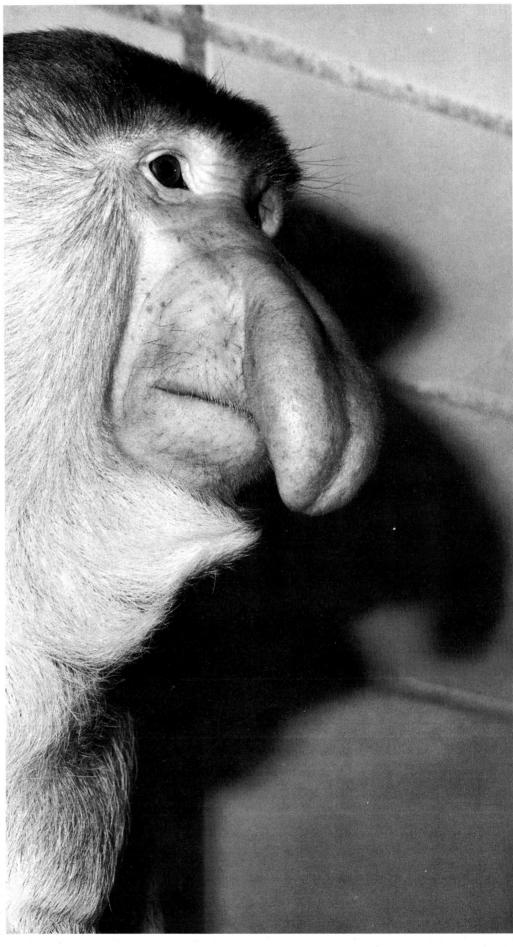

136–138 Proboscis Monkeys (*Nasalis larvatus*).
The face is dominated by the huge pendulous nose.
Only the young and female proboscis monkeys
have pointed turned-up noses. This makes it easy
to determine their sex.
Right: Stuttgart Zoological Gardens
Above and opposite page: West Berlin Zoological
Gardens

Opposite page (above left):
140 Northern Black and White Guereza
(Colobus abyssinicus).
No strict order of precedence within a colobus
troop seems to exist and consequently there
are hardly any fierce battles between rivals
for dominance. (Cologne Zoological Gardens)

Opposite page (below left):
141 *Colobus polykomos polykomos*.
There is not much difference in size between
this guereza and the northern variety;
the shorter fur, however, makes it look slimmer.
(Leipzig Zoological Gardens)

142/143 Red Colobus Monkeys *(Colobus badius).*
They spend most of their time in the middle
and higher regions of the tree cover where they
also eat and roam about. They rarely come down
to the ground. (Cologne Zoological Gardens)

144 Red Colobus Monkey *(Colobus badius)*.
Individual coloring varies from reddish brown
to black and their general appearance differs too.
The pinkish rings round the eyes, lips and
muzzle contrast with the normally gray face.
(Cologne Zoological Gardens)

145 Kilimanjaro Colobus Monkey
(Colobus abyssinicus caudatus).
The newly born have pure white, somewhat
curly fur which gradually turns darker when
the animal is between 6 and 10 weeks old.
(Dresden Zoological Gardens)

146 Kilimanjaro Colobus Monkey
(*Colobus abyssinicus caudatus*).
The male has an extremely long bushy tail.
(Dresden Zoological Gardens)

147 Southern Black and White Guereza
(*Colobus polykomos*).
There are sometimes light brownish tints
on the dorsal side of the black coat.
(Duisburg Zoological Gardens)

They move swiftly and nimbly along partly regular trails and paths through the dense vegetation of the virgin forest, whereby they frequently come down to the ground. Their diet consists chiefly of fruit, berries, foliage, young shoots and buds. These imposing animals are only found in zoological gardens with larger guenon collections. The two major subspecies inhabit disjunct areas:

Cercopithecus l'hoesti l'hoesti, eastern part of Zaire, Uganda, upland regions around Mt. Kilimanjaro; *Cercopithecus l'hoesti preussi*, northern Cameroun to Fernando Póo.

Diana Monkey *(Cercopithecus diana)*. In terms of coloring and marking the Diana monkeys belong to the most attractive guenons. The snowy almost gleaming white fur extending from the chin to the chest and along the front parts of the arms as well as the thin white stripes on the flanks form a distinct contrast to the dark gray body coloring. The hind-parts and inner surface of the flanks are chestnut brown to yellowish. The back is also brown. The length of the white beard varies according to subspecies. The Diana monkey's chin beard is shorter than that of the Roloway subspecies and is black in front. These guenons have a body length of up to 65 cm while the tail is even longer. They weigh between 6 and 9 kilos. With their jutting jaws the shape of their heads resembles that of the mangabeys. They are notable for their supple, feline movements; like the vervet monkeys the tail is often carried upright.

Diana monkeys are very lively; their curiosity or shyness is dependent on their mood and age. In captivity they are usually friendly animals. As the inhabitants of belts of tropical forest with tall trees they spend most of their time in the canopy. Despite their conspicuous markings they are difficult to discern from below. Contact within the groups is chiefly maintained by rasping sounds, chattering and long drawn-out cries. They largely subsist on fruit, seeds, nuts, young shoots, and buds.

Unfortunately the Diana monkeys too are hunted for their flesh and effective protective measures are therefore urgently needed. In zoological gardens with larger primate collections Diana monkeys are common and they breed well. Of the three subspecies the best known are:

Cercopithecus diana diana, Gambia to Liberia, and *Cercopithecus diana roloway*, Ivory Coast, Ghana. Also classified as a subspecies is *Cercopithecus dryas*, originally classed as a distinct species, which, however, occurs in the Congo region a great distance away from the other forms.

De Brazza's Monkey *(Cercopithecus neglectus)*. As far as its head markings are concerned, this guenon is quite as handsome as the Diana species. Its body measurements, too, are almost the same, though it is sometimes smaller and lighter. It has an orange diadem-shaped band across its forehead which stands out against the dark crown with its tufts of fur on the sides. The white nose, mouth-parts and beard form a similar contrast to the speckled light-brown grayish green color of the body. It has grayish blue "spectacles" round its eyes. A white stripe runs along the flanks which are dark in front.

The comparatively large area in which this species occurs extends from northern Cameroun via the Central African Republic and the forested southern part of Sudan to Kenya and southward to Shaba (formerly Katanga). De Brazza's monkeys inhabit both the higher altitudes as well as the plains, large belts of forest as well as strips of gallery forest and grove-like groups of trees, provided there is no open country without cover in between, and are also found on the fringe of savannahs with denser vegetation. Nor do they avoid swamps with tree cover. In some stretches of open country their territory touches on that of the vervet monkeys. As adept climbers they jump nimbly and surefootedly about the branches. They are equally mobile on the ground. At the approach of danger they bound away with upright tails—presumably a signaling device.

Besides fruit, young shoots, tubers, and roots these guenons are partly carnivorous, eating a large quantity of insects. In zoos De Brazza's monkeys are fairly hardy and remain amicable when kept in smaller groups from early youth and provided with spacious accommodation. As in the case of the Diana monkeys, breeding successes are fairly frequent, although usually only one young per group is born each year. Despite the extensive region in which they occur, De Brazza's monkeys have an identical appearance. It is therefore not necessary to list the five geographical subspecies.

Lesser White-nosed Guenon *(Cercopithecus petaurista)*. The forms (subspecies) dealt with under this heading are partly still classed in various species. They have a body length of 45 cm; the tail, however, may exceed 60 cm in length. Because they are so small they look like non-adult animals even when they have reached full maturity. The varieties differ conspicuously in their respective markings, particularly on the head. The white-nosed monkey is a uniform light grayish brown with olive-green tints. The lower sides of the head, the throat, chest, abdomen as well as the inner surface of the legs are white with a whitish gray tinge. The dark face is conspicuously marked with a pure white heart-shaped nose-spot, from which also their name derives.

---

148 Proboscis Monkeys *(Nasalis larvatus)*. With increasing age the pendulous nose, which may reach a length of 10 cm, droops over the mouth so that it has to be lifted or pushed aside with the hand when the animal eats. (Stuttgart Zoological Gardens)

All members of this group are well able to adapt themselves to varying conditions; although they give preference to different types of rain forest, they also inhabit swamps with tree cover, thickets on the fringe of woodlands, palm groves and the like. They can be found both in the upper and middle tiers of the tropical forest. For various reasons these very lively and mobile animals sometimes leave their original territory and settle in more distant regions. Each form possesses its own range which is usually separated from the others by natural barriers such as broad rivers.

The smallest social unit is the family with one male, one or two females and two or three offspring. A distinct order of precedence has not been observed. Resident groups occupy the same sleeping-places for a considerable period. So far territorial rivalries have not been noted. T. Struhsaker, who studied Schmidt's spot-nosed monkey in the Kibale forests (West Uganda), noted that groups numbering up to 35 animals were led by just one strong adult male. After an almost two-year "term of office" the dominant male was ousted by a rival who proceeded to kill two babies and then partly devoured them. Older children, however, were not attacked. Such behavior is probably an exception rather than the rule and is possibly a spontaneous aggressiveness unleashed by over-excitement at the time of achieving dominance. These guenons feed on fruit, nuts, young leaves, shoots, buds, and blossoms; they also make inroads on plantations when the opportunity presents itself. They eat insects too. The menstruation cycle lasts 30 days, gestation about seven months. White-nosed monkeys are often found in zoos with larger primate collections. The following subspecies are listed here:

Cercopithecus petaurista petaurista, West Africa, Ghana to Gambia; *Cercopithecus petaurista fantiensis*, West Africa; *Cercopithecus petaurista buettikoferi* West Africa, Sierra Leone, Liberia; *Cercopithecus petaurista stämpflii*, West Africa, Ivory Coast.

The following belong to *Cercopithecus ascanius*, a group that, though closely related, was formerly separately classified: *Cercopithecus petaurista ascanius*, West Africa, Lower Congo; *Cercopithecus petaurista schmidti*, blueish face, white beard, reddish brown tail, Upper Congo to Uganda. Five other geographical subspecies have been identified in the Congo region.

Greater White-nosed Guenon (*Cercopithecus nictitans*). Although the coat of this guenon is fairly dark all over, the individual black hairs have a fine yellowish to olive-green shading. Only the head reveals more or less distinguishing features, with a more oval-shaped pure white spot on the nose. Some of the twelve geographical subspecies that have been classified have whitish gray to white chins, throats and abdomens. Similar to the previous species, changes in the color of certain lighter parts of the coat are caused by specific substances in the diet.

The body length is from 50 to 65 cm, the very long tail may measure nearly 90 cm; they weigh between 7.5 and 9 kilos. These guenons occur over an extensive area from southern Senegal via Liberia to the eastern part of the Central African Republic, and western Uganda is also included by some authorities. As forest-dwellers they prefer high trees and spend most of their time in the tree-tops, only rarely coming down to the ground. It is said that within their range they remain inside specific territorial limits. The forest provides them with adequate fare the year round and their diet is similar to that of the lesser white-nosed monkey. This species, too, is fairly common in zoological gardens. The following belong to the three recognized but not exactly defined subspecies:

Cercopithecus nictitans nictitans, Cameroun and northern Gabon; *Cercopithecus nictitans martini*, West Africa, Benin to northwestern Cameroun (Sanaga river), Fernando Póo.

Red-bellied Guenon (*Cercopithecus erythrogaster*). The size and weight of this species is comparable with that of the larger white-nosed monkey varieties. The body length ranges between 50 and 60 cm and the tail is from 75 to 90 cm long. Formerly the red-bellied guenon was classified as a subspecies of the white-nosed monkey, with which they are apparently closely related. The fact that these monkeys only occur in southern Nigeria from Lagos to the lower course of the Niger and are isolated from all the other guenons justifies the status of a distinct species. These animals, too, have a conspicuous white spot on the nose. There are two types of coloring that occur frequently within the individual troops: animals with faded reddish brown chest and abdomen and those with gray to light brown fur on these parts of the body. In their way of life and habits there are no marked differences to those of their close relatives. In zoological gardens this species is only sometimes found in larger primate collections.

Moustached Guenon (*Cercopithecus cephus*). This guenon belongs to a group with conspicuous facial coloring. Varieties with reddish colored mouth and nose parts are sometimes classified as distinct species. Both in their size—the body length is from 40 to 60 cm, the tail from 60 to 80 cm long—and in their disposition these guenons closely resemble the lesser white-nosed monkeys. Characteristic features are the light blue horizontal stripe on the upper lip, naked blueish face, light-colored upper eyelids. Long whitish yellow whiskers grow on the cheeks. Like many other guenons the coloring is dark grayish brown with light brown tips to the fur and pale green tints. Throat, chest and abdomen are whitish gray. The tip of the tail is a faded reddish brown. These monkeys inhabit belts of tropical rain forest with dense vegetation, and rarely come down to the ground. They eat all kinds of fruit, buds, tree seed, berries, and oil palm pulp but they also consume more foliage than other species, such as the mona and greater white-nosed monkeys with which they sometimes associate as well as with mangabeys. They do not drink very much and quench their thirst with rain water from the trees and with the juice of the fruit they eat. They occur in an area extending from Cameroun (Sanaga river) to the region west of the lower course of the Congo (Gabon) where they form smaller groups of from four to nine animals. They belong to the more common species of the genus in zoological gardens since they do not require specialized care.

Five subspecies have been named although there are still differing views about their nomenclature. They have been classified into two groups, to which the **red-nosed** and **Sclater's monkeys** also belong.

Hamlyn's Monkey (*Cercopithecus hamlyni*). Also known as the owl-faced guenon, this species has some unusual features. With an overall length of 110 or 120 cm, its tail is about 60 cm long. It weighs between 6 and 8 kilos. It has a white stripe along the ridge of its hooked nose which, seen from the front, really looks like the pointed beak of a large type of owl. The owl-like impression is enhanced by the shape of the face and eyes and the long fur which falls outward. The conspicuous white stripe along the nose is an optical signaling device by which members of the species can recognize each other in the dim light of the dense tropical forest. These visual signals are of immense importance in helping the group to keep together especially since the coat camouflages the animals' presence in dense foliage. When the monkey is agitated or alarmed, the fur round the face

stands on end and the animal nods its head up and down and sometimes to and fro in a similar manner to owls when they are in danger or disturbed. This threatening gesture is a specific instinctive reflex. There are still many gaps in our knowledge about these monkeys' habits.

They occur in an area north of the Congo and along the Ituri river on the territory of Zaire. The exact limits of their range have still to be defined. They move along regular "trails" in the upper regions of the dense rain forest. As adept clamberers and leapers, whereby the tail has an aerodynamic function as in the case of many other guenon species, they roam nimbly and swiftly in small troops over their territory.

They are mainly vegetarians, insects and other small animals merely constituting supplementary fare. They begin to reproduce when they are about five years old. After gestation lasting about six and a half to seven months, one baby is born which begins to move independently among the branches when it is only a few weeks old. Until then it is carried clinging ventrally to its mother. It later maintains loose contact with her until the start of puberty at the end of its fourth year. Owl-faced guenons are rare animals and are only found in a few zoological gardens. Inadequate knowledge makes it impossible to name the subspecies.

Talapoin Monkey (*Cercopithecus talapoin*). With a body length of only 32 to 37 cm, a tail about 40 cm long and weighing from 1.3 to 1.6 kilos, the talapoin is the smallest species of Old World monkeys. The relatively large head with its domed forehead and the flat mouth-parts recall the Lorenz child scheme. The

Swamp monkey
(*Cercopithecus nigroviridis*)

female differs from the other species in that she has large swellings on the outer genitals when she is sexually receptive. The teeth, too, differ in some respects to those of related species. The talapoin monkey has noticeably short fingers connected by webbed skin at the base. In addition, its repertory of sounds is not similar to that of other guenons. In short, it deviates to such an extent from the other members of its genus that several zoologists have classified it as a distinct genus or subgenus *(Miopithecus)*. It should, however, be borne in mind that it closely resembles a guenon in other respects, such as its brownish gray body coloring with greenish tints as well as the whitish to light grey underside. Its face, which is adorned with a short moustache and yellowish beard, is largely naked and of a faded pink color.

It occurs in three separate West African regions extending from southwestern Cameroun to Angola. A more limited range also exists in the Ruwenzori Mountains on the border between Zaire and Uganda, where the talapoin monkeys are mainly found in the higher arboreal regions of swamps and also in mangrove zones on the coast. Their troops number from about 30 to 60 animals although they split up into smaller units for foraging and feeding, forming larger groups for periods of rest and travel and at their common sleeping places. They associate temporarily with mona, greater white-nosed and moustached monkeys.

The size of the troops makes it hard to identify a definite order of precedence. While on the move one of the males quite frequently utters a loud call which helps the inconspicuously marked animals to keep together when they cannot see each other. When at rest several animals often sit close together, loosely entwining their drooping tails.

Talapoin monkeys are largely vegetarians, with preference given to fruit and palm nuts. Additional fare is taken in the form of insects and other small vertebrates. The talapoins are not common zoo animals, although they have known to breed there. There are four geographical subspecies which have so far not been clearly defined. Generally speaking though, they correspond with the previously named four areas of distribution: southwestern Cameroun and Gabon; near the lower course of the Congo; in Angola as well as in the Ruwenzori region.

Swamp Monkey *(Cercopithecus nigroviridis)*. In the early eighteen-nineties a guenon with dark fur was exhibited at the London Zoo. It was, however, not subjected to an exact determination. In 1907, thirteen years after the death of the animal, which had been preserved and stuffed, the British zoologist Pocock identified it as a new species. As a result of some anatomical features which revealed a certain resemblance to those of the mangabeys and baboons—in the main concerning the structure of the skull and teeth—some zoologists classified it as a full genus *(Allenopithecus)*.

With a body length of about 45 to 48 cm and a tail ca. 50 cm long, these monkeys belong to the smaller guenon species. The blackish body fur has yellowish green tips which gives it its Latin name. Other specific features are a black beard, pinkish gray rings round the eyes, yellowish white throat, chest, abdomen and inner surface of the limbs and long fair hair on the sides of the neck. These guenons live in the damp forests of the central and northwestern Congo regions that extend to Gabon. Little is known so far about their life when resident in the wild. They deviate both in their habits and calls from other species. From observation made in zoos it seems that small troops of them enjoy splashing about in shallow water as they probably do when the forests are temporarily flooded. They are apparently of a very lively nature. Their diet is composed mainly of various fruit, unripe nuts, buds, and leaves. There is no exact information about whether they also eat food of animal origin. Since only a few swamp monkeys have so far been exported, they are very rarely seen in zoos.

Patas Monkeys (genus *Erythrocebus*)

Patas or **Red Monkey** *(Erythrocebus patas)*. In adaptation to a life on the ground the patas monkeys have developed a number of conspicuous features in their physical proportions and behavior so that they are classed as a distinct genus. In open grassland they keep to a certain order of march like soldiers, thus maintaining contact and giving the group protection. The body length ranges between 60 and 87 cm, the tail is from 50 to 75 cm long. The males, which have very fierce-looking canines, weigh up to 24 kilos, the smaller females being much lighter. Due to very long and slender arms and legs—a typical example of adaptation to life in wide expanses of open country—the male has a shoulder height of almost 60 cm. Hands and feet are shorter than those of the other guenons. When running, the heels and wrists are partially raised from the ground. In open spaces, especially when in flight, they may reach a speed of 50 kilometers per hour whereby they can also make sudden turns. According to the zoologist Hall they may be rated as the "greyhounds" among the primates.

Their territory covers a maximum of 52 square kilometers and they travel daily distances ranging from one to eleven kilometers. Apart from this they make relatively little use of their running ability. In the high grass they are hidden from view and, in contrast to other guenons, are hardly ever heard. Only the young animals at play race about. As inhabitants of the open grasslands they avoid the forest and even small groups of trees. They are, of course, capable of climbing trees, but seldom do so. The young animals are clumsy and accident-prone clamberers. Only one fully adult and outstandingly large male holds sway over the herd which as a rule numbers between 5 and 25 animals—18 on average. He is primarily responsible for guarding and protecting the group from all external dangers. He keeps watch, constantly on the alert. When acute danger threatens the females seek safety in flight without keeping any sort of order. The male attempts to dupe the enemy by creating a diversion: he displays himself on a bush or higher ground, or diverts attention by deliberately fleeing in the wrong direction. The patas' enemies are jackals, hyenas, and sometimes leopards, gepards or crowned eagles.

The females have an order of precedence that is rarely infringed. They are headed by an "alpha" female who is even respected by the males and whose offspring, too, enjoy privileges. Should males attack them, the females put up a determined joint resistance. The males are sexually mature at an age of three and a half to four years. The dominant male drives these potential rivals out of the group. When resting, the status of the individual female can be discerned by her sitting distance to the dominant male.

The patas' diet consists of various plants as well as insects and small animals, such as lizards. Their range that partly overlaps with that of the vervet monkey and baboons, extends over the dry grasslands to the north of the West African belt of tropical forest to the Sudan and northern Kenya in the east. To the north the vast desert zone forms a natural barrier. Patas monkeys are common in zoological gardens where their chief requirement is spacious accommodation. Groups kept for breeding purposes nearly always include young animals. There are two subspecies:

Erythrocebus patas patas, West Africa—Senegal, Niger, and Chad regions; *Erythrocebus patas pyrrhonotus*, East Africa—Sudan, Kenya.

Leaf Monkeys (family Colobidae)

The leaf monkeys are still sometimes grouped together with the guenons; they are classified here as a distinct family. Although the genera differ substantially from each other in appearance, they are largely identical in their anatomical structure. Despite their slender build some species weigh up to 23 kilos and have a body length of up to 82 cm. The head is rounder than that of the guenons; the thumbs are relatively short or even reduced to a stump (colobus monkey); the legs are generally longer than the arms; with one exception (Pageh pig-tailed langur) the tail is very long; they have no cheek pouches. As their name indicates, these monkeys live almost entirely on the foliage of various trees. Their dental structure and shape of the teeth are functionally adapted to this diet. The back teeth, furnished with transverse crests, also move backward and forward when chewing so that the leaves are pulverized. The large stomach which,

when full, occupies almost a third of the body volume (about 20 percent of the body weight), is composed of three chambers, with a gullet running into the middle one. This physical resemblance to the ruminants' stomach induced Bauchop and Martucci (1968) and Kuhn (1964) to make a closer study of the digestive system of colobus monkeys and langurs respectively. In the first two chambers of the stomach a rich brew of bacteria attacks the cellulose in a similar way to that of the true ruminants. No trace of protozoa was found in the langur stomach. Although containing plenty of fibrous substance, their diet of leaves is not very nutritious and large quantities have to be consumed to sustain the animal's energy. The intestines are correspondingly large. When their stomachs are full these animals therefore need lengthy rests to digest their food. After eating they often sit, apparently idle, at their chosen resting-places.

The leaf monkeys are classified into six genera: the langurs, douc and snub-nosed monkeys, Pageh pig-tailed langurs and proboscis monkeys live in South Asia while the colobus monkeys are found in Africa. The South Asian species in particular have adapted themselves to a wide variety of living conditions: damp hot jungles, mangrove swamps, open cultivated country, arid zones and cool upland forests. Now that more is known about their biological needs, leaf monkeys are more commonly seen in zoological gardens.

Langurs (genus *Presbytis*)

To a certain extent the langurs embody the prototype of the manifold forms of Asia's leaf monkeys. Notable features are their slim build, long limbs, long thin hands, relatively small thumbs, a very long tail, a large larynx sac that acts as a resonator and usually a monochrome coat with lighter fur on the ventral side: the head is often adorned with a crest, comb, crown, beard, etc. The color of the newborn's fur often differs from that of the adult animal. The langurs eat their food, which partly consists of fruit and seeds, both on the ground and in the trees. When running fast over the ground they have a dog-like gait and move about the trees by clambering, hanging by their arms as well as by taking long leaps. Small groups of powerful males sometimes acquire a harem by force and this leads to a division of the herd. There is hardly any rivalry when they are sympatric with the rhesus monkeys since the latter as omnivores are not competitors for food. When traveling, the dominant male usually heads the group. Their most dangerous enemy is the leopard; tigers however seldom attack them. At the approach of danger loud alarm calls are sounded which cause the entire group to disperse in headlong flight. There are four subgenera.

Swinging its arms forward
the hanuman monkey *(Presbytis entellus)*
leaps to the next tree providing sustenance.

Langurs (subgenus *Semnopithecus*)

Hanuman Monkey *(Presbytis entellus)*. Also known as the common langur, Hindi langur or entellus monkey. With a head and body length of up to 82 cm and a tail that is about one meter long in older males, the hanuman monkey is the largest langur. The coat is usually a light medium gray and may vary locally. The naked parts of the face as well as the hands and feet are black. Older animals have large bulges over their eyes and long, black bristling eyebrows which jut out like the peak of a cap. The babies have a darker grayish brown coat. Hanuman monkeys usually hold their tails upright or drooping over their backs. Although mainly leaf-eaters, they also feed on herbs, grass stalks, soft bark and roots, blossoms, rice, euphorbia, berries, corn-cobs, fruit, and resin. As sacred monkeys they are fed well and regularly by the Hindus in and outside the temples. Tolerated, protected and fed by human beings, hanuman monkeys are often very bold and not infrequently steal food from the table.

Accounts of their habits in the wild have been given by Phyllis Jay (1965), Ripley (1970) and Vogel (1976) among others. The hanuman monkeys probably have the most variable social structure of all the primates. This is to a certain extent probably an effect of their ability to adapt themselves to a broad spectrum of ecological conditions. Since they are quite accustomed to the presence of human beings in many places they spend more time on the ground than other monkeys, their ability to clamber nimbly and leap distances of up to ten meters enabling them to seek safety in flight should danger threaten. The size of the groups varies a great deal—from two to about two hundred animals: the average size is between four and forty. On the edge of the jungle only small harems have been observed. Led by a male these defend their territory. Large herds have many males who usually occupy various positions, which are subject to frequent change, in the order of precedence. The highest-ranking, so-called alpha animal enjoys full privileges while the other males form a loosely-knit "men's association." Large groups soon tend to social instability and stagnation so that some of the males and females form one or more new subsidiary units or one-male harems. This does not always happen amicably. Male intruders may force their way into the herd, oust the dominant males and subjugate the harem group. In order to consolidate and demonstrate his power a new alpha animal sometimes kills the smaller children and expels the male adolescents; the females are taken as mates. Such actions, of course, verge on disturbed behavior. It may also happen that strong solitary males join the herd, accompanying it for a time and then leaving it with a group of females. Individual males, too, sometimes change groups; troops of males that have reached adulthood in isolation may also attack the harem males in the group and successfully oust them. According to observations made, for instance, in South India, this kind of behavior is engendered by growing population pressure in a confined territory.

In its social function the community life of the hanuman monkeys deviates from that of the baboons and macaques where mutual protection is the prime consideration. The hanuman society's main purpose is the rearing of offspring and the integration of the young animals into the community system. The birth of a baby elevates mother and child to a privileged status irrespective of order of precedence. All the females display great interest and take a part in rearing the baby, even looking after it temporarily. The fathers almost completely ignore the children. The development of the offspring takes place in several phases: infancy (dark coat) lasts up to the fourth or sixth month; after weaning, from about the tenth to twelfth month, comes the age of playing and learning; the adolescent phase begins for the females in their third and for the males in their fourth year; integration into the order of precedence is a gradual process. Whereas the young females remain more with their mothers and often occupy themselves with their younger brothers and sisters, the young and still playful males tend to seek the proximity of the older males on the fringe of the group until they are driven off or leave the group of their own accord. They begin to reproduce when they are about five and a half to six years old. Gestation lasts about five and a half months. In some places the majority of births occur from April to June but otherwise newly born are found the whole year round.

Conflicts between several hanuman groups—as well as those within the group—generally end without serious injuries. Aggressive acts in these cases seem to be governed by instinctive ceremonial rules and deaths rarely result. Ritualized "battles," which may be engendered by lack of living space, occur sometimes in South India between hanuman groups and also within the individual groups, the size of whose territory ranges from 0.5 to 7.8 square kilometers. Contact calls between groups, even at a greater distance, are often heard. In zones where the nights are cold the hanuman monkeys' daily routine begins with a warming up procedure.

In Indian mythology, too, the hanuman monkey played a prominent part. In the *Ramayana* and in legends and sagas he symbolized the self-sacrificing loyalty of a friend. Pious Hindus worship these monkeys as an embodiment of Hanuman, the monkey god, who, in the shape of Sugriva, liberated Sita, the divine consort of Rama—a personification of the god Vishnu—after she had been abducted to Sri Lanka (Ceylon) by the giant Ravan. While rescuing her, Sugriva discovered the prized mango fruit and brought it back as booty to India. For this crime he was condemned to be burned at the stake but managed to extinguish the fire in time although his hands, feet and face were scorched. Since then all hanuman monkeys have black feet, hands and faces.

Hanuman monkeys occur all over the Indian subcontinent from the Indus to the Brahmaputra, from Kashmir to Cape Comorin. Up to thirty years ago the life expectancy of langurs in captivity was very low although the hanuman monkey with its somewhat better ability to adapt was kept with relative success in our latitudes. It was extremely complicated to supply these animals with their specialized diet of green foliage in winter. Digestive upsets were the main cause of gastro-enteric ailments which ultimately led to death. Recent advances in nutritional science have now made it possible to achieve better survival and breeding results with the aid of a diet that is biologically adequate to their specific needs. The Indian zoologist S.H. Prater has identified 15 subspecies of which the following are named here:

Presbytis entellus entellus, light-gray coat, northern part of the Indian subcontinent with the exception of very dry zones; *Presbytis entellus schistaceus*, largest form with a broad beard and longer thick coat of a lighter color, southern Himalayas from the upper Indus valley (Kashmir). In summer up to 4,000 meters altitude in rhododendron bushes and cedar woods and sometimes above the timber-line. Has even been observed on the snow-line thus becoming associated with the Yeti, the "abominable snow man." *Presbytis entellus hypoleucus*, feet and tail with clear black markings, west coast of India; *Presbytis entellus priam*, crest-like comb of black fur, Southeast India and Sri Lanka.

Purple-faced Langurs (subgenus *Kasi*)

John's Langur *(Presbytis johni)*. Its distinguishing features are a sleek black coat and a light golden brown head, bewhiskered on the cheeks and chin. The long black bristles of the jutting eyebrows stand out in contrast. The face is of a faded brownish color with a faint tinge of purple. Older males may have a body length of up to 70 cm, the tail measuring up to 80 cm. This animal's relatively small range is located in southwestern India extending from the Nilgiri Mountains and the Cardamom range to the southern slopes of the Western Ghats in altitudes ranging from 800 to 2,100 meters. Like all langurs it mainly feeds on leaves, buds, and in certain places, acacia twigs and blossoms. When alarmed it moves very swiftly with long bounds, reaching a speed of from 30 to 40 kilometers per hour. The size of the groups may range from three to 25 animals, the average being about nine. The territory of a group covers from 0.6 to 2.6 square kilometers. In some places the population density even exceeds 100 animals per square kilometer (Poirier 1968); but usually it is much lower, especially where the herds have been depleted by local non-Hindu tribes that hunt them both for their flesh and fur. The flesh, blood and certain organs of John's langur are said to have curative properties. Hindus who are less strict in the practice of their religion also hunt them.

Not very much is known about the life of John's langurs in the wild. One strong male leads the troop, a third of which may consist of males. Some of them may leave the group or are expelled. Similar to the hanuman monkeys, although more rarely, ritualized battles are said to occur when the territory is confined. The groups keep in touch over longer distances with contact calls.

Less care would appear to be lavished on the offspring. In exceptional cases mothers have been seen to temporarily leave their barely month-old babies clinging to twigs, ignoring their cries. The newly born differ in their coloring to the older animals; they are a golden brown color similar to that of the dusky leaf monkey. According to Poirier (1968) these langurs have a proper baby-sitting system—a female looks after three or four babies whilst the other females and mothers go foraging. An important result of this habit is the acquisition of social behavior by the children whilst the females who look after them, particularly the adolescents who handle the babies clumsily and roughly to begin with, develop their maternal abilities. In zoological gardens John's langurs are still very rare and sensitive charges. They are registered by the IUCN as an endangered species.

Purple-faced Langur *(Presbytis senex)*. Like John's langur it is lighter and more slender in build than the hanuman monkey. Its local name "wandhura" has caused it to be mistakenly named wanderoo, the designation of the lion-tailed monkey which is a macaque. The coat varies from a silvery gray to dark brown or grayish black; the head, however, is always of a brownish hue. The broad white beard forms a distinct contrast to the dark face. These langurs occur in Sri Lanka where several subspecies have been identified. In Sri Lanka it enjoys the same status as the hanuman monkeys on the mainland. As a result it is found not only in forests but also in close proximity to human settlements where it is tolerated and protected. In some places it is hunted for its flesh and fur. On the move and especially when fleeing from an enemy, it is a swift runner. It prefers an arboreal life to that on the ground. Contact calls between individual groups, ritualized battles in confined territory as well as biting bouts—the latter occurring more frequently than among other species—have been noted. In their eating habits these monkeys are very adaptable. It was observed, for instance, that

a number of them foraged for food on the rocky coast in a region where all the trees had been felled and only low thickets remained which they used as sleeping places. Otherwise they have no remarkable habits. Except for Indian zoological gardens these monkeys are rarely exhibited since they are very difficult to keep.

Capped Langurs (subgenus *Trachypithecus*)

In Farther India the capped langurs occupy the same position as the hanuman monkeys in the southern part of the subcontinent. The numerous forms which occur in separate habitats extending over Assam, Bangladesh, South China, northern Vietnam, Sumatera, Kalimantan and Java almost defy exact classification. In course of time many geographical subspecies have evolved in a zone where changes in the earth's crust have split a land mass into fragmented morphological regions and islands. In recent years the former classification into twelve species with 27 subspecies has been superseded by a grouping into merely seven species, but views still differ widely in this respect. The fact that color varieties occur frequently within the individual species adds to the difficulties of making a clear definition. Typical features are various crest-like growths of hair on the head, brows that are less jutting and a sleeker coat.

Silvered Leaf Monkey *(Presbytis cristatus)*. Its coat is medium to dark brown with an occasional reddish tinge. A notable feature is the growth of long hair, on the sides of the face too, which falls backward partly covering the ears, the fur on the head resembling a kind of black cap. These monkeys are powerfully built with a body length of 70 or 80 cm. The individual groups number from 20 to 51 animals, with an average size of about 30. The territory of a single group is said to cover only 0.2 square kilometers, and under favorable conditions 150 animals can exist on just one square kilometer. There is always a majority of females in the groups. Individual males may form harems within the group. When conflicts occur or the population pressure within and between the groups increases, they not infrequently transfer themselves to another troop. Nevertheless, ritualized battles and confrontations are fairly uncommon. Their extensive range extends over Farther India south of the 12th degree of latitude to Sumatera and Kalimantan, one of the subspecies even occurring on Java. These langurs, too, are regarded as sacred animals in some places, but in others are hunted for their flesh by the local people. Silvered leaf monkeys are uncommon in zoos but are occasionally exhibited in South and Southeast Asia. Breeding successes have been frequently recorded. Several subspecies have been identified:

Presbytis cristatus cristatus, Southeast Asian mainland; *Presbytis cristatus germaini*, Sumatera, Kalimantan; *Presbytis cris-*

tatus pyrrhus, Java and Bali. Formerly classified as a distinct species, it occupies a special position, being smaller than the other subspecies with a very dark to blackish coat flecked with sparse white hairs. Crest and cap are absent but the face is heavily bearded. Soft fur with a silky sheen.

Capped Langur (*Presbytis pileatus*). The fur coloring of this species varies considerably according to the geographical region. The northern varieties have a medium gray to light brownish tinge on the upper side of the body. Abdomen, chest, sides of the head as well as the inner surface of the limbs, on the other hand, are light rusty-brown to faded orange; the face is dark gray to black. The crest is formed by long dark brown upstanding fur. Other varieties sometimes have a lighter coloring, especially on the back. The southern varieties with a body length of about 50 to 60 cm are smaller. As far as is known their habits correspond to those of the other langurs. Their range extends from Assam via Upper Burma to Yunnan (South China). Like *Presbytis cristatus pyrrhus* they used to be fairly common in zoological gardens. About 50 years ago a specimen lived eight years in London, and another one 23 years in San Diego, which probably corresponds to their natural life expectancy. The following subspecies are named as examples:

Presbytis pileatus pileatus, Assam; *Presbytis pileatus durga*, large dark crest and long brownish hair on the sides of the face. Largest form. Assam; *Presbytis pileatus geei*, also classified as a distinct species. Has a thick cream to orange colored coat. Very long hair on the sides of the head. Black face. Registered by the IUCN as rare. Border region of Bhutan to the east of the Sankosh river; only occurs in small numbers. In the region extending from Assam to Upper Burma the forms *Presbytis pileatus brahma* and *Presbytis pileatus tenebricus* occur.

Phayre's Leaf Monkey (*Presbytis phayrei*). Although it closely resembles the capped langur there is no dark pigment in the skin round the eyes which thus have a whitish to flesh-colored appearance. The face is therefore similar to that of the dusky leaf monkey. No detailed information is available about its habits. It occurs from Bangladesh and South Assam via Burma and Thailand to the south of Vietnam.

The stomach, intestines and gall bladder of the capped langurs in particular often contain lumps of calcium carbonate or fur which have been swallowed and pressed together to form hard balls almost the size of a chicken's egg. Known as "stomach stones" or "bezoar balls" they are believed to possess miraculous curative powers according to Indian and Chinese medical popular beliefs especially in the case of the Bengal langur. In consequence these animals are hunted in large numbers. Phayre's leaf monkeys are uncommon in zoological gardens.

Of the three subspecies *Presbytis phayrei barbei* has a smaller crest and is very rare; Bangladesh, southwestern Assam.

Dusky Leaf Monkey (*Presbytis obscurus*). Its coat is grayish blue or blackish according to variety; the underside is lighter. The tail is sometimes a faded yellowish gray. The white rings round the eyes are a conspicuous feature. The skin between mouth and nose and on the lower lip also has white markings. The cap of hair on its head is less conspicuous. According to sex the body measures between 55 and 65 cm while the tail may be up to 80 cm long; the weight is said to be 11 or 12 kilos. This species occurs from southern Thailand, Tenasserim via the Malay Peninsula southward to Johore (Singapore) and the adjoining islands. The offspring of the dark-colored langur species have a light orange-hued coat during their first months of life. When they are three months old the coat begins to assume the color of the adult animal. Development in childhood and adolescence is similar to that of the hanuman monkeys. The size of the groups ranges from 9 to 17 animals, with an average of 13. These leaf monkeys are common in zoos with larger primate collections.

The subspecies closely resemble each other. The best-known is: *Presbytis obscurus flavicauda*, Malay Peninsula.

Francois' Leaf Monkey (*Presbytis francoisi*). Little is known about the habits of this and the following capped langur species. Francois' monkey has conspicuous white stripes from the corner of the mouth to the ear and a large crest with two tufts. The body is mainly black. Some forms have light to golden brown or yellow tops to their heads according to variety; this coloring is sometimes seen on the neck too. This species occurs in the northern parts of Laos and Vietnam.

So far these monkeys have been rarely exhibited in zoos since they are considered to be fastidious charges. The subspecies are still a matter of dispute. Sanderson and Steinbacher (1957) mention as species:

Presbytis (Trachypithecus) laotum in Laos; *Presbytis (Trachypithecus) delacouri* in Central Vietnam; and *Presbytis (Trachypithecus) poliocephalus* in Laos and northern Vietnam. Except for minor coloring variations these animals all look very alike. They could therefore justifiably be listed here as subspecies. The classification of *Presbytis (Trachypithecus) shanicus*, a langur identified by the same authors as occurring in the Shan states, remains open here. It could possibly be classified as a subspecies of Phayre's leaf monkey.

White-headed Leaf Monkey (*Presbytis leucocephalus*). There is still no general agreement about the exact classification of this species. So far the region where it occurs has hardly been studied in terms of its zoology. It was first discovered in 1966 by Chinese zoologists in the mountainous region of Kwangsi, South China. The white shoulders, head and feet contrast with the dark coat. Almost nothing is known about its habits. There are no records of its having been kept in captivity. Closely related to Francois' monkey, some zoologists classify it as a subspecies.

Mentawai Island Langur *(Presbytis potenzani)*. It is black on the upper side of its body, the ventral side being a bright yellow to golden orange, while the cheeks and throat are white. It comes from a small group of islands off the west coast of Sumatera, a considerable distance away from the region where the other capped langurs occur. It lives in small troops or family parties and also in pairs. Like the dusky leaf monkey the newborn is golden orange in coloring and thus matches the fur on the mother's ventral side. There is no information about this species in captivity.

Island Langurs (subgenus *Presbytis*)

They almost equal their cousins on the mainland in the variability of their forms. Smaller than the capped langurs, the head and body length ranges between 45 and 60 cm, while the tail is from 50 to 70 cm long; their approximate weight is from about 8 to 12 kilos. They are distributed over the islands of Southeast Asia. Small troops—rarely more than 15 animals—permanently inhabit specific wooded areas, their territorial habits being similar to those of the guenons. They are often not in the least shy and supplement their diet with what they find in gardens and plantations. In addition to leaves they also eat ripe and unripe fruit. Sanderson (1967) observed that they sometimes eat food of animal origin. Although these langurs are rare in captivity, it has been established in zoos that gestation lasts about 140 days, a shorter period than that of the other langurs. In Indonesia these species are still regarded as sacred animals with a special status in religion and mythology. Only a few of the variously classified species and subspecies are listed here.

Maroon Leaf Monkey *(Presbytis rubicundus)*. The broad spectrum of color varieties, which make an exact classification very difficult, ranges from a monochrome reddish brown to a red coat with a faded whitish tinge. There is a large crest of fur on the head, the face is dark gray. The newborn are conspicuously light to white in coloring. They are very uncommon even in Southeast Asian zoos. Amongst the subspecies are:
Presbytis rubicundus rubicundus, reddish to chestnut brown. Southern parts of Kalimantan; *Presbytis rubicundus carimatae*, pale to reddish brown. Abdomen and face whitish gray. Northern parts of Kalimantan.

Banded Leaf Monkey *(Presbytis melalophus)*. Sometimes called the black-headed langur and also named simpai in its own country. Back, fore limbs, feet and upper surface of the tail are dark brown to black, the rest of the body pale to medium reddish brown. In contrast, the beard on the cheeks and chin is yellowish white. The head is crowned by a transverse crest. Bernstein (1967) recorded some observations about the habits

of these monkeys in the wild. In the Malayan forest they live communally with dusky leaf monkeys, pig-tailed macaques, Javanese monkeys and white-handed gibbons. Each species maintains a certain distance from the neighboring group of its own species. Apart from leaves, these langurs eat a fairly large amount of fruit and small quantities of other vegetarian food. Several zoos in recent years have reported increasing successes with these animals. They occur from the Malay Peninsula to Sumatera. A few of the subspecies:
Presbytis melalophus melalophus, southern part of the Malay Peninsula, western Sumatera; *Presbytis melalophus chrysomelas*, one of the varieties is almost entirely black; *Presbytis cruciger*, formerly classified as a species, is nowadays regarded as a color variety of this subspecies.

White-fronted Leaf Monkey *(Presbytis frontatus)*. The body is a brownish gray color while the limbs are black. The fur on the head forms two tufts in front of the cap. The naked skin on the forehead has a lighter coloring. Only found on Kalimantan.

Sunda Island Leaf Monkey *(Presbytis aygula)*. A very beautiful species that occurs on Java, Kalimantan and also Sumatera where it is named "lutong." The dark gray fur on its back contrasts with the white coat extending from the chin to the tail on the underside. Behind the tuft on the forehead, which is white with a thin vertical black stripe, there is a thick, high, dark crest. The cheeks are surrounded by a prolific growth of white upturned whiskers which form an attractive contrast to the dark gray to black face.

Like the hanuman monkey in India, the lutongs on Sumatera are regarded by the Bataks in the mountainous regions as sacred and are still worshipped. On Java, too, this monkey is often seen in close proximity to villages and in gardens and plantations. Sanderson (1967) observed them eating large tree snails. This species was the only leaf monkey known to Linnaeus and he included it in his *Systema naturae*. These fastidious leaf monkeys are chiefly found in zoos situated in their range, for instance in Jakarta. Different face markings made it possible to identify the individual subspecies.

The best-known subspecies is *Presbytis aygula thomasi*, with a black stripe on the white forehead and a pink upper lip. Java, Sumatera.

According to Sanderson (1957) there are also the following species, listed here to round off the picture:
Presbytis femoralis, Tenasserim (Malay Peninsula) over Sumatera to the west coast of Kalimantan; *Presbytis hosei*, Kalimantan in the uplands to an altitude of over 1,200 meters; *Presbytis sabana*, northern part of Kalimantan. Based on Napier and Napier (1967). A. Jolly (1975) does not classify these forms as species.

Douc Langurs (genus *Pygathrix*)

Although closely related to the langurs the douc langurs are classified as a distinct genus with only one species because of certain features of their skull and unusual fur markings.

Douc Langur *(Pygathrix nemaeus)*. These animals are sturdily built but do not have such long limbs as the langurs. Adult males have a head and body length of 60 or 70 cm; the females are somewhat smaller. The French naturalist Buffon was the first to describe their distinctive coloring. The upper part of the head is gray to brownish; two orange stripes link the outer corners of the eyes and ears; the neck has a whitish yellow and a chestnut brown band; body and upper part of the arms are gray; flanks, hands and feet are black. The lower part of the arms is white, the lower legs are reddish brown. A white patch on the rump and a white tail heighten the effect of the color contrasts. The short-haired thick fur is beautifully glossy.

These monkeys live almost entirely on leaves. One baby is born at a time. Since the females are also sexually receptive during pregnancy and mate during this time, exact information about gestation is not available.

This species is distributed over the tropical rain forests of Vietnam, Laos and Hainan Island. Registered by the IUCN as an endangered species, these animals have become more common in zoological gardens in recent years. Two subspecies have been identified:

Pygathrix nemaeus nemaeus, with a pale face, southern Vietnam, southern Laos. *Pygathrix nemaeus nigripes,* with a dark face, northern Vietnam, northern Laos.

Golden monkey
(Rhinopithecus roxellanae)

Snub-nosed and Proboscis Monkeys

(genera *Rhinopithecus, Simias, Nasalis*)

According to W. Fiedler and H. Wendt (1967) the three distinct genera can be regarded as a single complex with the possible status of subgenera. Within the leaf monkey family these primates are easily distinguishable. Their diet of leaves is highly specialized. Typical features are the oddly shaped "humanlike" noses, the structure of the skull and the foot which indicate transitional phases within the three genera.

Snub-nosed Monkeys (genus *Rhinopithecus*)

Compared with the previously mentioned species these monkeys are more thickset in build and the skull appears more massive; the upper arms are relatively long. Both sexes possess the typical upturned nose. Inhabiting upland forests they have largely adapted to low temperatures and a damper environ-

Pageh pig-tailed langurs *(Simias concolor)*

ment, their coat being very long and thick. The stomach is particularly voluminous. Two to four species have been classified.

Golden Monkey *(Rhinopithecus roxellanae)*. On Ancient Chinese vases dating from about 2200 B.C. golden, bright-colored, human-like creatures with extremely grotesque turned-up noses can be seen. Up to about a century ago it was believed that these figures represented fantastic fabulous beings. Then the Jesuit priest Armand David acquired in the Chinese-Tibetan borderlands a monkey skin that resembled these representations and brought it back to Paris.

The coat of this beautiful animal is a dark slate to brownish color from the top of the head to the tail on the back including the flanks. Head, neck, chest, abdomen, legs and the inner surface of the arms are a glowing yellow orange. The naked face is grayish brown. With a body measuring from about 55 to 78 cm and a tail that is between 60 to 87 cm long, this species is one of the largest amongst the snub-nosed monkeys. They occur over an extensive area from eastern Upper Burma via Yunnan to Szechwan at an altitude of between 2,000 and 3,000 meters. In this inaccessible zone it is very difficult to discover and observe them. In large troops they travel swinging by their arms and leaping from branch to branch through the coniferous mountain forests and bamboo thickets, feeding on leaves, buds, bamboo shoots, and fruit. Although they descend to the valleys

in winter, these hardy animals have also been observed in snow-clad forests. Their thick winter coat is exchanged for a thinner one in summer.

In bygone days all golden monkey hides had to be surrendered against payment to the Chinese imperial authorities. Because they are so very rare and difficult to obtain they have never been exhibited in Europe and America. The Peking and Shanghai zoos succeeded in acclimatizing two groups of golden monkeys during the nineteen-sixties and they have repeatedly bred. Registered by the IUCN as rare animals. The classification of the related varieties as species or subspecies of the golden monkey is still in dispute.

Brown Snub-nosed Monkey *(Rhinopithecus bieti)*. With a body length of about 80 cm this species equals or even exceeds the golden monkey's measurements. Forehead, head, shoulders, back, tail, flanks as well as arms and legs are brownish black, the inner surface of the limbs reddish brown which merges into a grayish white on the upper part of the arms. Chest and abdomen are also grayish white; the light blue rings round the eyes contrast with the gray mouth-parts. This monkey occurs in the mountain forests of Yunnan (South China). It is considered a very rare species. There is no information about the existence of these monkeys in zoos. As already mentioned, they are possibly a subspecies of the golden monkey.

White-mantled Snub-nosed Monkey *(Rhinopithecus brelichi)*. Its back is gray to brown and the longer white fur on the shoulders gives the impression of a cape. Arms and crest are yellowish. This monkey is very large, maximum head and body length measuring about 85 cm. It occurs in the upland forests of Kweichow extending into the northern part of Szechwan. There is no information about its existence in zoos. It is also regarded as a subspecies of the golden monkey.

Tonkin Snub-nosed Monkey *(Rhinopithecus avunculus)*. This species has some specific anatomical features which distinguish it from the other snub-nosed monkeys. It was previously classified as a distinct genus *(Presbyticus)*. This classification, however, is not universally accepted; it is therefore included here as a species.

With a head and body length of between 50 and 55 cm and a tail about 65 cm long, this monkey is definitely smaller than its close relatives. The dorsal side and limbs are black, the ventral side whitish yellow, the front part of the head pale brown to yellow, the back of the head and nape orange to brown. Hands and feet are short and broad. The pointed nose is markedly turned up. It is found in the mountain forests of northern Vietnam. There is no information about its existence in captivity outside Asia.

Pageh Pig-tailed Monkeys (genus *Simias*)

Pageh Pig-tailed Monkey *(Simias concolor)*. The genus Simias consists of only one species. In recent years R. Tilson has succeeded in throwing light on the previously almost unknown existence of this monkey. It only occurs on the islands of Siberut, Sipora and South Pagai which belong to the Mentawai Islands off the western shores of Sumatera.

The Pageh pig-tailed monkey has a head and body length of about 50 or 55 cm but its naked tail is only 15 cm long with sparse hairs on its tip. The structure of the skull is similar to that of the snub-nosed monkeys. Two color varieties exist: one is dark gray to black with narrow white bands on its forehead and similar cheeks and brows; the other is lighter in color, being of a golden yellow to brownish hue with a white forehead and whitish colored lower arms and legs. The face of both varieties is black. The snub-nose is relatively small.

These shy monkeys live in the dense rain forest and normally do not often come down to the ground. In pairs or family groups with two or three offspring they travel through their territory in search of special kinds of leaves, buds, blossoms and fruit. The chronic meat shortage has caused the local people to hunt these monkeys for their flesh. This and large-scale forest clearance pose a threat to the future survival of this already very rare species. Pageh pig-tailed monkeys have so far not been exhibited in zoos. Sanderson expresses the view that several races or subspecies exist on each of the islands.

Proboscis Monkeys (genus *Nasalis*)

Proboscis Monkey *(Nasalis larvatus)*. There is really nothing odd about the fact that the Malays named the proboscis monkey "orang blanda"—"white man." A certain facial resemblance cannot be denied. This animal is altogether something of an oddity amongst the leaf monkeys.

It is big and sturdily built. The male's head to body length varies between 66 and 76 cm, the female measuring only from 54 to 60 cm, and the tail is from 56 to 70 cm long. Strong males may weigh up to about 22 kilos. Only the children and young females have snub noses. The males, on the other hand, when they are sexually mature at about six years and even more frequently with increasing age, develop a long pendulous nose. Whether this secondary sex characteristic is merely a signaling device with symbolic significance has not yet been determined. It possibly also acts as a nasal resonator. Adult females can be identified by their forward-facing pointed nose. The coat is a deep to pale reddish brown, head and back are a darker shade than the ventral side. Proboscis monkeys only occur on Kalimantan in wooded areas in close proximity to water, including rivers and lakes, but mostly in the mangrove swamps on the coast and near the mouths of rivers, where they are found in groups of up to about 20 animals, including several adult males. They move deliberately among the branches, but react swiftly if necessary, leaping a distance of over four meters horizontally or jumping downward over eight meters with spread-eagled arms, legs, and tail. They are able swimmers—crossing even broad rivers—and can avoid danger by jumping into the water and also remaining under the surface for up to 28 seconds. In the heat of the midday sun they keep cool by paddling in shallow water.

Compared with the langurs, their social system is less rigidly organized. To a greater degree than most other primates the proboscis monkey depends on a specialized diet for sustenance, including sonneratia leaves and nipa palm blossoms—typical vegetation of the mangrove swamps. According to Kern (1964) about 95 percent of their diet consists of this fare. In addition they eat various buds, shoots, bamboo tips as well as fruit containing bitter substances. At Frankfurt on Main and Cologne Zoos the proboscis monkeys relish flour grubs and locusts so it can be assumed that they also sometimes eat food of animal origin when living in the wild. They spend most of the day searching for food. Adequate digestion of their voluminous diet of leaves takes a long time. Some observers say that, like the ruminants, the proboscis monkeys bring up the mash from their stomachs and chew it again.

Only one young is born at a time and its face is a blueish color to begin with. On the basis of field studies, Sanderson (1957) gives a rough estimate of about 166 days for the gestation period and an approximate weight of 500 grams at birth. Like the douc monkeys the pairs stay together during the female's pregnancy, and mating continues, so that the exact start of gestation remains problematical. To a certain extent the males help to rear the young and join in the play of adolescent offspring. The crocodile is one of the proboscis monkeys' natural enemies. The Dajaks hunt them for their flesh.

The first satisfactory results in rearing and breeding these animals in captivity were scored in the early nineteen-sixties. Around the turn of the century the zoo in Cairo acquired this fastidious monkey, followed by the London Zoo in 1902. A proboscis monkey in the San Diego Zoo survived from 1956 to 1960—a record period—and in 1965 another of these monkeys was born and reared there, a zoological sensation since it was the first to be born in captivity outside Indonesia. During recent years other more or less successful breeding efforts have been recorded. The IUCN has registered it as a rare species. Current experience seems to indicate that zoo successes with these fastidious animals are not just dependent on a specific diet. Factors of equal importance are biologically adequate substitute fare, continuous veterinary supervision and the best possible accommodation. Given these conditions, animals that were in good health at the time of their arrival acclimatized themselves quickly and took to the new diet: apart from large quantities of oak, willow, robinia, hazel-nut, rose, blackberry, raspberry, mulberry, and other leaves, they were given fruit in season, carrots, tomatoes, cucumbers, potatoes, rice, corn-cobs, and other vegetables. The two subspecies are:

Nasalis larvatus larvatus, western Kalimantan; *Nasalis larvatus orientalis*, with a somewhat lighter coat, eastern Kalimantan.

Colobus Monkeys (genus *Colobus*)

"Kolobus" is a Greek word meaning mutilated. The name refers to the stump-like rudiments of the thumb which, however, is normal in the embryo. In all probability an initial stage of hand-over-hand arboreal locomotion was the cause of this. In build the colobus monkeys are shorter and more thickset than the langurs. The head and body length ranges between 50 and 75 cm, the tail is from 60 to 100 cm long and they weigh from 10 to 12.5 kilos; the legs are longer than the arms. Some of the species have long fringes of hair on their coats.

Their specialized diet has already been described in connection with the leaf monkeys. In the case of the exclusively vegetarian colobus monkeys their intake of vegetable protein, with the exception of some essential amino-acids, provides an adequate substitute for animal protein. The bacteria in the stomach that are digested while breaking up the cellulose probably adjust the protein balance. On the other hand, monkeys that principally live on fruit lacking in proteins generally eat larger quantities of animal fare that is rich in proteins.

The leaf monkey probably originally evolved in Asia. Today's colobus monkeys migrated to Africa via Southwest Asia and Southeast Europe, as Pliocene fossil remains indicate. Today they live isolated from each other on two continents. Colobus monkeys are distributed across Central Africa from Senegal and Angola in the west to Ethiopia and Tanzania in the east, with the exception of arid treeless zones. Their range is the tropical rain forest where they are mainly found in the middle tiers and tree-tops. They move through the labyrinth of branches, running, swinging from bough to bough by their arms and leaping. As semi-brachiators they demonstrate an early form of swinging beneath the branches. Figures given by W. B. Collins with respect to a 250 square kilometer area of a forest reservation in Ghana indicate that locally the colobus monkeys are more numerous than all the other monkeys put together. Their principal natural enemy is the crowned eagle.

Like the guenons and langurs, the numerous forms of colobus monkeys, which nowadays are all registered as rare animals, have been classified in a more condensed form. They can be divided into three groups which according to Napier and Napier (1967) may also be given the status of subgenera:

Olive colobus monkey (*Colobus verus*), subgenus *Procolobus*, **red colobus monkey** (*Colobus badius*), subgenus *Piliocolobus*, and **Guereza** (*Colobus* spec.), subgenus *Colobus*.

Colobus Monkeys (subgenus *Procolobus*)

Olive Colobus Monkey (*Colobus verus*). With a head and body length of 50 cm and a tail that is 65 cm long this species is the smallest and probably the most ancestral type of colobus monkey. Its head, too, is relatively small. Its olive-green coloring is less vivid, it has no notable crest, and its digestive organs are less specialized in their structure. These factors entitle it to a special status. The face is naked or only sparsely haired. More adaptable than other species, it inhabits various kinds of forest from deciduous dry woodlands to damp swampy and rain forest in a zone extending from Sierra Leone to western Nigeria. Not very much is known about its habits when living in the wild. Less socially minded than other species, it forms groups numbering from 5 to 20 animals—from 10 to 15 on average—and is sometimes found living in pairs. It prefers to live in the middle and lower regions of the forest, sleeping and finding sustenance there. The females develop distinct genital swellings even before they attain sexual maturity. Like the mouse lemur, the mother carries her newborn in her mouth. A few weeks after birth it starts to cling to her fairly short fur. The species is very uncommon in zoological gardens.

Colobus Monkeys (subgenus *Piliocolobus*)

Red Colobus Monkey *(Colobus badius)*. Its head and body length is between 60 and 75 cm, the tail may be up to 95 cm long, the weight ranges between 8 and 11.5 kilos. The head appears relatively small; the tail may be long or short haired and is without a tuft. Little information is available about its habits. The red colobus monkey spends most of its time in the middle regions or high tree-tops of the forest and finds its specialized diet on trees that are just in leaf, since this fresh foliage contains more proteins and glucose. Red colobus monkeys roam their territory in large groups numbering from about 30 to 150 animals.

They occur throughout the forest belts of West and Central Africa as well as in some small disjunct regions of Kenya and Tanzania. These animals' natural habitat is severely threatened by growing human settlement and resulting forest clearance. Groups of red colobus monkeys that were observed near the Tana river (Kenya) averaged about 18 individuals of which never more than two and often only one were adult males. In larger tracts of forest the groups number from 36 to 40 animals. Usually the females give birth to one baby only every two years, irrespective of season. The reserves established in recent years on the lower course of the Tana river with its gallery forests afford protection for the red colobus monkey, the agile mangabey and other animals.

In some parts of West Africa these monkeys used to be regarded as messengers of the gods. Like the langurs in India, they bask in the warmth of the first and last sunshine of the day. The red colobus monkeys are very uncommon in zoological gardens. The following are some of the numerous subspecies:

Colobus badius rufomitratus, head reddish brown, body darker. East Africa and eastern Zaire; *Colobus badius gordonorum*, central part of East Africa; *Colobus badius tholloni*, West Africa, Gabon, Cameroun; *Colobus badius ferrugineus*, dorsal side and upper part of the head blackish brown, neck, arms and legs rust red. West Africa; *Colobus badius ellioti*, ventral side rust red, dorsal side and legs darker. West Africa; *Colobus badius kirkii* is classified by Napier and Napier (1967) as a distinct species. It has conspicuous markings, the back and tail being reddish brown, shoulders and arms blackish brown, ventral side whitish gray, the dark face surrounded by a whitish ruff. Very rare and threatened with extinction. On Zanzibar there are only 200 animals and attempts have been made to transfer them to the mainland.

Black and White Colobus Monkeys
or **Guerezas** (subgenus *Colobus*)

The many forms and varieties make it hard to establish a uniform classification. There are two distinct species: the northern black and white guereza *(Colobus abyssinicus)* and the southern black and white guereza *(Colobus polykomos)*. Like the previous species they have hard patches on their hindquarters but these are often covered by fur so that they can hardly be seen. To begin with, the baby is carried in its mother's arms. But after a few days it clings to the mother's abdominal and shoulder fur, thus hardly hampering her in her movements. When they are about seven to eight weeks old the offspring begin to eat leaves. They are not weaned, however, until they are seven or eight months old. At an early age the young animals attempt to go short distances from the ever watchful mother and to climb. They learn to be independent early on. The babies are often carried around by other females. Generally speaking the baby is given back to the mother if it cries with fright or is hungry. This behavior is probably not just a product of the confined conditions in captivity.

Guereza children are very playful. The father, too, patiently lets them play with him. The guerezas begin to reproduce when they are about five years old; gestation lasts ca. six months. Zoo records indicate that they live to an age of about 25 years, though their life expectancy is probably shorter when living in the wild. Here, too, the diet largely consists of leaves, and according to local observations they also relish tender *Rauwolfia* twigs and shoots, as well as *Podocarpus* and *Juniperus procera* berries. Guereza stomachs have always been found completely filled with food weighing up to 3 kilos—a third or a quarter of the adult animal's weight. Guerezas rarely descend to the ground.

Northern Black and White Guereza *(Colobus abyssinicus, Colobus guereza)*. The guereza mountain varieties in particular have long white fringes along their bodies and fan-like tufts on their tails which make them appear larger than they really are. In profile the flattened tip of the nose in the dark gray to black finely haired face is a characteristic feature. The body itself has a thick, deep black glossy coat. This form lives mainly on the slopes of Mt. Kilimanjaro and Mt. Meru at altitudes from 1,000 to 3,000 meters. In between there are varieties with different kinds of ruffs and crests on their heads. The guerezas that occur in Uganda, for instance, only have a very thin shoulder mane; in Kenya and Ethiopia they have more or less large tufts on their tails. The black and white patterned fur is naturally an easy means of identification for mutual recognition amongst the animals of a group, but for the observer on the ground the body outline is indiscernible in the interplay of light and shadow in the labyrinth of branches. When disturbed, these monkeys remain motionless at first. If danger becomes acute, they utter alarm calls and flee, displaying their nimbleness in clambering and prowess as long jumpers. Downward leaps of up to 12 meters are nothing exceptional. When taking off the guerezas first extend their legs and arms forward, then bend them briefly, stretching them again just before landing elastically in the branches. In downward leaps the outstretched legs and arms and flattering shoulder mane produce a winged effect that breaks the fall, while the bushy tail acts as a steering or balancing mechanism. They appear to glide through space.

According to W. Ullrich, who made a close study of guerezas in the Meru region, these animals usually remain in the same locality, and the territory of the individual family groups is also fixed. Even when fleeing they rarely leave their own territory which covers approximately 10 to 18 hectares, but which may vary according to the size of the group—about 6 to 20 animals with an average of from 8 to 12—and the available food supply. As a rule, at certain times of the day the animals take the same routes and this is connected with their search for food. The marking of territory and maintenance of contact within the group is conducted vocally. When an encounter between several groups takes place the males usually "bark" briefly and repeatedly; ritualized battles rarely take place and not many injuries are inflicted when they do. The males threaten an enemy by violently shaking the twigs and scampering noisily amongst the branches. A strong male acts as leader of the group and guarantees the comparatively loose order of precedence. He tries to ward off intruders by threatening behavior—stiffening his legs and demonstratively chewing—and covers the usually disorderly retreat of the group when danger is in the offing.

Sexually receptive females are taken as mates by any of the males, but the bond is only of brief duration. Despite the lack of a thumb guerezas too engage in mutual grooming as a social activity.

The handsome coat, in particular of the mountain guerezas, attracted the attention of men early on. There was a trade in these hides in the Roman Empire and Marco Polo reported that the rulers of Central Asia paid high prices for them. Various African tribes, including the Massai, used them for ornamental purposes. Around the turn of the 20th century hundreds of thousands of guerezas were sacrificed to the dictates of European and American fashion. In 1892 alone 175,000 hides were supplied to the European market so that the total figure must have run into millions. Thanks to the vagaries of fashion the rage for monkey fur was short-lived and the guerezas escaped total extinction. In recent decades too, the growth of tourism has been accompanied by a mounting demand for these hides as souvenirs. However, the majority of the depleted herds have now returned to normal strength and this is also due to protective measures, particularly in the national parks and reserves. This species is distributed from the mostly wet forest of the plains and the mountainous regions of Ethiopia, Kenya, and northern Tanzania in the east over southern Sudan, the Central African Republic, Cameroun to eastern Nigeria in the west.

With better knowledge about the habits of these monkeys and the introduction of more modern zoo techniques—including the use of deep-freeze foliage in winter—the life expectancy of guerezas in captivity has considerably improved and they now survive for much longer and breed. Hence they have become fairly common in larger zoological gardens. Of the many subspecies which differ in color markings and fur embellishments the following are listed here:

Colobus abyssinicus abyssinicus, with a white tuft on its tail—Ethiopia, Kenya, eastern Uganda; *Colobus abyssinicus caudatus*, with a long fringe of fur on the shoulders and loins, long and very bushy tail. Called "mbega" in Swahili by the Massais. Kenya, northern Tanzania; *Colobus abyssinicus kikuyuensis* and *Colobus abyssinicus palliatus*.

Southern Black and White Guereza (*Colobus polykomos*). In its basic coloring it differs from the Abyssinian species by being entirely black with variations according to the subspecies. The fringe of hair on the flanks is either sparse in growth, patchy or is entirely lacking, in particular amongst the West African forms. The markings on the face also differ, as well as the fur on the normally white tail which only has a very modest tuft.

The little-known habits of this guereza probably do not differ much from those of the other type. It has a more southerly distribution, adjoining that of the northern guereza and extending to northern Angola and southern Tanzania to about 10 degrees southern latitude. The forest extending from Gambia to Togo/Benin forms an isolated range. This guereza is not as common as the other type in zoos. A few of the various subspecies are:

Colobus polykomos polykomos, with a black face surrounded by a thick white ruff, grayish white fringe on the flanks—West Africa; *Colobus polykomos angolensis*, with a wide white ruff on the sides of the head, black forehead, tufts of white hair on the shoulders, black body with a dark grayish tinge on the back. The base of the white tail is black—Angola, Congo Basin, Shaba, Malawi; *Colobus polykomos satanas*, entirely black, its mane partly hangs over its face—West Africa, Cameroun, Fernando Póo. *Colobus polykomos vellerosus* is very similar to *Colobus polykomos*—West Africa.

Apes and Men
(superfamily Hominoidea)

The apes are our nearest relatives in the animal kingdom. The gradual transition to bipedalism, leaving the arms and hands free for manipulation, as well as a communication medium that was equally essential for social life, resulted in the use of language, the development of the ability to generalize and, ultimately, to thought in its highest form. The constant interplay of these basic factors engendered continuous intellectual and social evolution. Man has become a social and creatively productive being. He nevertheless remains a constituent part of nature and in this role dominates it to an increasing extent, tapping its resources as the basis of his existence.

Today not only biologists and zoologists, but also anthropologists, archaeologists, physicians, veterinary experts, psychologists, philosophers and, last but not least, social scientists study the relevant aspects of these most highly developed primates. The closer we come to the evolutionary components of the human race, the more inevitable are comparisons with our own "species." The major criteria are anatomy, physiology, morphology, serology, ontology, ecology and behaviorism. Comparative studies help us to discover common roots and parallels in the structure and functions of the body and its organs. But even when we are aware of our affinity with nature we cannot eliminate emotional aspects of a specifically human or individual character that impinge upon our acceptance of the fact that we are directly involved in this evolutionary process. With the descent of men and apes from common ancestors, from subhuman primates, we embody this genealogical heritage both in original and adapted forms. Whatever views people may hold, this affects them directly. The manifold aspects of the question are, however, too complex to be dealt with fully in a biological description of the primates. This brief characterization of the new evolutionary quality represented by *Homo sapiens* is merely intended as a general outline.

Gibbons (family Hylobatidae)

The lineage of the gibbons was for a long time the subject of controversies. Proceeding from the gibbon-like apes in the mid-Tertiary period which at that time occurred both in Europe and Africa and were not as specialized as they are today, the conclusion was reached that these "pre-gibbons" formed a link between the Old World monkeys and the apes. On the basis of the more human-like features of existent gibbons, such as standing upright on two legs, the astounding resemblance of the embryo gibbon to human beings—the long arms do not de-

velop ontogenetically until a very late stage—Ernst Haeckel regarded them as our nearest kin amongst the primates. Nonetheless the gibbons differ substantially from the other apes. They still have hard patches on their hindquarters which, although smaller, resemble those of the Old World monkeys, and their brain is less convoluted. Although the gibbons are classified as apes belonging to a distinct or independent group, approximately 25 to 35 million years ago they began to follow a very different line of development to that which ultimately produced human beings, becoming highly specialized in swinging along beneath the branches by their arms. As a result today's gibbons do not belong to the human family tree. This fact is reflected in the modern classification of gibbons as a distinct family instead of a subfamily. As an adapted primary means of locomotion their arms are indeed very long and powerful, their full stretch exceeding that of the legs and torso combined. A man's arms only reach a third of the way down his thighs; those of an upright gibbon touch his ankles. On the ground the gibbon runs or trips along holding its slightly bent arms either sideways or over its head, the hands being cupped like hooks. It holds its arms in the same way when it nimbly runs along a horizontal branch, using them elegantly as balancing poles. The gibbons are the only highly developed primates to move on two legs when they come down to the ground. Swinging by their arms beneath the branches is simply a reversal of bipedalism and is known as brachiation. Gibbons are true forest-dwellers and could not exist without trees. Moving almost effortlessly at high speed, at times reaching almost 50 kilometers per hour, they travel through the tree-tops, hurling themselves ten or twelve meters across space to the next tree, or swooping downward for an equal distance in a slanting glide. With extended hands and feet they find a sure hold as soon as they land elastically on a branch. They either clamber hand over hand, swing by their arms beneath or between the branches or jump forwards and downward, often so rapidly that it is almost impossible to follow the separate movements. They grab the branches so briefly that they appear to fly rather than swing. Impressed by this glorious sight the poets Zhang Chao and Meng Chao wrote 1100 years ago: "He sits in the clouds. He sails with the wind and reaches out for the moon." The gibbon is, in fact, an accomplished acrobat.

Like the other apes the gibbons have no tail. The comparatively small head, which is sometimes adorned with characteristic tufts, makes the body appear larger. A sagittal crest on the cranium to which the larger temporal muscles are attached is absent. The long fang-like canine teeth, with which deep wounds can be inflicted, are intimidating. The head and body length ranges from 45 to 90 cm according to species. The color of the coat varies considerably within the species too. There are animals with pale, whitish, grayish yellow to brownish or even dark to black coats respectively. The color of the thick and soft fur may change in the course of the animal's life. The differently patterned white markings on the face are characteristic for each species—with the exception of the siamangs. In adaptation to an almost entirely arboreal life the lower arms with the extremely long, slim but powerful hands have become exceptionally elongated and muscular. In comparison with the hand of the orang utan with its short and weak thumb the gibbon's thumb, although it has moved down towards the wrist, is long and can be used when clambering if necessary. The long fingers have developed into hooks that are used to obtain a firm hold. The astounding pliability of the shoulder joint is necessary for swinging beneath the branches or orientation in a hanging position. It enables the gibbon to turn through more than 180° in all directions. The gibbon also owes its astonishing agility and apparent defiance of the laws of gravity to its lack of weight, which, with the exception of the siamang, rarely exceeds 10 kilos. The arms have a stretch of between 130 and 180 cm. All gibbons sing in loud characteristic choruses which can be heard for a great distance and almost certainly serve as a vocal marking of territory, as orientation and indication of their presence for individual animals too. Usually these choruses form part of the daily routine and can be heard every morning and before the brief twilight begins to fall. The various species can be more easily distinguished by their calls than by their appearance. The males and females each sing differently and there is even individual variation so that this complicated structure of sounds indicates the species, sex, age, and the identity of the individual vocalist. They generally return to the same trees to sleep but in contrast to the large apes do not build platforms. The size of the group's territory is largely dependent on its numbers, fertility rate and the amount of food available; it ranges from 12 to 120 hectares, the average size being from 30 to 40 hectares. This territory is defended against intruders that are warned to keep away by alarm calls and, if necessary, defensive actions. An attack on alien animals may occur at lightning speed. Always on the alert, they heed even the slightest unusual sound. When danger threatens they retreat at full speed.

Gibbons only form family groups, each consisting of the parents accompanied by their offspring which are born at intervals of two to three years. These groups number five or six animals at the maximum. Normally a gibbon leaves the family at the age of 6 or 7 years, when it is ready to mate. They live monogamously and the pairs usually remain together for the rest of their lives. Mating often occurs while the male and female each hang by one arm from a branch entwining their free arms and legs. Gestation takes about 210 or 212 days. Toward the end of the second and start of the third month of pregnancy the female's vulva becomes swollen—an important sign for zoo staff. According to observations made in zoos, when giving birth the mother uses her hands to prevent the baby from falling. The baby is lovingly cared for and carried around by its mother during its first months of life, clinging to the fur of her abdomen. The offspring are fully mobile at the age of ten months although they do not attempt big leaps at this age.

The gibbons' diet is largely vegetarian, consisting of mango berries, unripe seeds, tender leaves, young shoots, buds, blossoms, bamboo tips; they also eat tree frogs, insects, and smaller birds, which they even catch on the wing, as well as eggs.

The gibbons are widely distributed over Southeast Asia to the Greater Sunda Islands. In Buddhist regions these apes are treated almost like sacred animals and in some places the local people keep them tamed or semi-tamed in their settlements where they are well treated. Various tribes, such as the Meo and Karen, hunt them. The gibbons are the most successful group of highly developed primates. According to current estimates they number about 200,000—far more than the herds of chimpanzees, gorillas, and orang utans. Nowadays their chief enemy is man—principally because of his inroads on their habitat. Sometimes young gibbons are caught after the mother has been shot, and sold on the black market. Their enemies in the animal kingdom are leopards, clouded leopards, Charsa martens, and more rarely snakes. A certain number of insignificant losses are also sustained as a result of parasites, illness and accidents such as falls involving fractured bones. The various species are generally classified as belonging to one of two genera.

Siamangs (genus *Symphalangus*)

Siamang (*Symphalangus syndactylus*). The siamang is by far the largest species of the gibbon family. Standing erect it is almost one meter high. Its arms have a stretch of nearly double this length. With a maximum weight of 20 kilos the male is nearly twice as heavy as the other species. The females are somewhat smaller. The difference between arm and body length is not quite so large as in the case of the true gibbons. The coat remains black at all ages; only the eyebrows have a brownish tinge. The long fur looks somewhat shaggy and untidy. An unusual feature is the membrane linking the second and third fingers for half their length. The siamang's loud voice is produced by a resonating larynx sac which inflates almost to the size of a baby's head when in use. At dawn and sunset, when agitated or confronted with an unusual situation, the siamangs' complicated chorus with its set sequence of calls, whereby male and female have a different vocal pitch and motif and a repertory of five different kinds of sound, can be heard for miles. Sometimes they hold their hands in front of their mouths when singing. It is thought that this is intended to cloak or dispel the intimidating effect on their partner of a wide open mouth that utters loud calls. With the aid of these noisy competitions the individual family groups vocally mark their territory and consequently the places where they forage, sleep and rest. The noisier groups usually assert themselves successfully against the less loud. The duets sung by male and female help to stabilize the pairs. Because of their piercing voices the siamangs are sometimes called the "howler apes" of the Old World.

They are distributed over the tropical rain forests of the Malay Peninsula and Sumatera up to an altitude of 2,000 meters. Their size and weight prevents their being as nimble as their smaller relations. But as opposed to the gibbons, they are fairly good swimmers. There are still considerable gaps in our knowledge about their habits when living in the wild. They live in pairs in family groups of up to six individuals. According to Chivers the male helps to rear the offspring. At night he sleeps with the children; the female with the baby. During the daytime the male often carries the youngest child.

Siamangs are sometimes found in larger zoos. Isolated breeding successes have also been recorded. Life expectancy in captivity now averages 20 years. The Grant Park Zoo in Atlanta (USA) caused quite a sensation with the two births in 1975 and 1976 of hybrids sired by a white-handed gibbon father with a siamang mother. The offspring stayed alive and resemble their mother. They have no larynx sac. Whether these hybrids can reproduce is not yet known. They are in any case very rare examples of cross-breeding.

Subspecies have not been established, although Sanderson mentions a smaller form which is said to live in the uplands of Selangor (Malaysia).

Dwarf Siamang (*Symphalangus klossi*). The classification of this species, first discovered in 1903, has not yet been finally clarified. The dwarf siamang is probably more closely related to the genus *Hylobates*, and in particular to the black gibbon. Its fur is very soft, sleek and has a handsome silky gloss. This makes the animal appear more slender. It is approximately the same size as the true gibbons. The male may weigh up to 8 kilos. The dwarf siamang has a smaller larynx sac. Hardly anything is known about its specific habits in the wild.

It occurs only on the small islands of Siberut and South Pagai of the Mentawai group adjoining the western coast of Sumatera. Although the dwarf siamangs are rarely hunted by the local people because they resemble human beings and their death is supposed to bring misfortune, the existing small herds have been sadly depleted. The intensive felling of timber threatens to destroy their entire habitat. In an effort to halt this process the Indonesian nature protection authorities placed 65 square kilometers of original rain forest under protection in 1974 as a reservation for the dwarf siamang, the Pageh pig-tailed monkey, the Mentawai leaf monkey and other rare animals. This Teitei Batti Reservation is situated in the Saibi river valley in the heart of Siberut. One can only hope that with the support of the World Wildlife Fund the rescue action will succeed. The species is threatened with extinction. Very uncommon in zoos.

True Gibbons (genus *Hylobates*)

Black Gibbon (*Hylobates concolor*). The true gibbon species are much more graceful and elegant in appearance than the larger and more robust siamangs. Their identification is sometimes extremely complicated. As previously mentioned, the color of the coat is subject to many changes. The coat of a young animal may change color later on or remain the same all its life. Some animals change color several times. Often male and female of the same species have differently colored coats. Hence in a single family group there are often several varieties of coloring. The gibbons' "chorus" sounds more mellow and melodious and is not so shrill as that of the siamangs.

In contrast to the other species the black gibbon has a larynx sac that is, however, much smaller than that of the siamang. The male has a crest of long fur that stands up along the middle of the head. The female has a small tuft of fur on each side of the head. Generally speaking, three forms or colour groupings that tally with their geographical distribution are identifiable and perhaps should be regarded as subspecies:

1) Both sexes are entirely black or dark brown as a rule, coastal region of northern Vietnam and on Hainan;

2) Male black and female yellow to dark brown with black on the top of the head; interior of northern Vietnam;

3) Male black with white cheeks; about 40 percent of the females are said to be a pale yellow color; from the western part of northern Vietnam, Laos, eastern Thailand to southern Vietnam. The latter form has already been classified as a subspecies —*Hylobates concolor leucogenys*.

In zoological gardens the black gibbon, along with the white-handed gibbon, is fairly common. In recent years black gibbons have repeatedly been bred in captivity.

Hoolock Gibbon (*Hylobates hoolock*). This species belongs to the smaller type of true gibbons. A conspicuous feature is the pale, normally white band just over the eyes that is usually interrupted by a narrow stripe above the nose. The males are always deep black, the females, often depending on their age, have various colorings from black to brownish or gray. The baby's coat is mainly gray. The fur is long and thick, hiding the hindquarter patches. Although smaller, the hoolock's voice is almost as loud as that of its larger relations. Its name is derived from its two-syllable call. Healthy hoolocks may live for more than 30 years; an animal in Lucknow Zoo (India) lived to an age of over 32 years.

The northwestern distribution border of hoolocks and indeed all gibbons is formed by the southern Himalayas in Bhutan. Towards the southeast they occur in Assam, Burma, Thailand, the western part of Laos and Yunnan.

Hoolock gibbons are uncommon in zoos outside their natural habitat.

White-handed Gibbon or **Lar** *(Hylobates lar)*. The lar is probably the best-known gibbon. Its coat varies from a black to buff color. A typical feature is the pale, usually white ruff that frames the face. As the name indicates, the hands and feet are white on the upper surface. The females are often lighter in color. Older males that were originally black sometimes turn medium brown later on, whereas buff-colored females become darker as they grow older. These gibbons occur in southern Burma, Tenasserim, southwestern Thailand, in the south of the Malay Peninsula and on Sumatera. We owe most of our knowledge about the gibbons' habits in the wild to the American primatologist Carpenter who was the first to make a special study of the white-handed gibbons. He established that young gibbons experience five stages of development:

1) Continuous contact with the mother, infancy, first coordinated movements.

2) Increased independence from the mother in locomotion; full set of milk teeth, still suckled, begin to play with each other, already utter alarm calls.

3) After two years they begin their adolescence; more independent of the mother, who may already have a new baby, vocal organs largely developed, clambering, swinging, playing.

4) Completely independent, very lively and playful, not very proficient in big leaps in which they sometimes receive support.

5) Final adolescent phase; in their habits still clearly recognizable as young animals, they display a pronounced urge for combative play; the teeth are still small.

In the phase of young adulthood there is very little to distinguish them from the older animals. Their canine teeth have not yet reached their full size. The nipples of both sexes are still small. The play instinct is only noticeable in a few instances. They attempt to dominate their younger brothers and sisters. They begin to loosen their ties with the family group at the age of about 6 years and this bond is severed when they are approximately 7 years and have started to become reproductive.

According to my own observations the first stage of the inevitable separation from the mother just before the birth of the next baby is a decisive phase; all her care is lavished on the newly born. Apparently bewildered and upset, tearful and very restless, the older child constantly seeks close contact with its mother, but is persistently rejected. Substitute contacts with its father or older brothers and sisters do not seem to satisfy it as yet. It takes about one or two weeks before the child regains its psychical stability. This crucial phase, of course, helps it to become more independent.

The successful acclimatization of these gibbons to zoo conditions over the past 30 years has resulted in many births. When reared as tame domestic pets they are extremely trusting and affectionate, as noted by W. Fischer. They are very adaptable and easy to teach, keep themselves clean and include the humans who look after them in the grooming procedure. The close social contact that these domesticated animals seek of their own accord, as well as their affectionate nature, is most probably based on the strong social bonds existing within their small family groups. In such cases the human being is regarded as family kin for a lengthy period. When adult, the tame gibbon only accepts those members of the family whom it implicitly trusts and even here sometimes behaves differently according to the sex and age of the human beings in the home. Everyone else is regarded as an intruder, attacked and quite often bitten. There are two subspecies:

Hylobates lar lar, nominate form, Lower Burma via Tenasserim to Sumatera (western form); and *Hylobates lar pileatus*, light-colored animals which have a black cap and a black patch on the chest. Some zoologists regard it as a distinct species. Registered by the IUCN as endangered. Southeast Thailand to Kampuchea (eastern form).

Agile Gibbon *(Hylobates agilis)*. Closely related to the lar and silvery gibbon. Since its coat is not so long and thick, the hindquarter patches are clearly visible. Agile gibbons vary considerably in their body coloring and face markings, ranging from black to a lighter or darker brown hue. But varieties with lighter limbs and darker bodies as well as vice versa are also known. Like the lar, the face may be framed by a ruff of paler fur.

In its habits the agile gibbon largely resembles the other species. It, too, remains the whole year round within the boundaries of its territory. The males which belong to neighboring families challenge each other with swift leaps; they rarely engage in biting bouts. In the course of these sham battles the females sometimes give their mates "moral" support by uttering loud screams.

149–151 Black Gibbons *(Hylobates concolor)*. Some of the females have a very light buff coat with a glossy sheen. The only contrasting colors are the dark or black patches on the top of the head and round the eyes. (Hanover Zoological Gardens)

Overleaf:

All five gibbon species have fairly similar habits. True gibbons are very poor swimmers. Water saturates their fur so quickly that acute danger of drowning exists. So gibbons avoid stretches of water and can move about freely in zoos when they are kept on islands surrounded by shallow channels. (Hanover Zoological Gardens)

152 Siamangs *(Symphalangus syndactylus)*.
Their voice is amplified by a conspicuously
large larynx sac which can be inflated like
a balloon. (Dresden Zoological Gardens,
photo: J. Berndt)

153 Black Gibbon *(Hylobates concolor)*.
When swinging by her arms the mother draws
her flanks up to the abdomen, thus forming
a kind of "nest" to protect her baby.
(Hanover Zoological Gardens)

154/155 White-handed Gibbons or **Lars**
(Hylobates lar).
Young gibbons develop relatively fast.
Females and babies have coats of either black
or light brown fur. The white hands and
feet are clearly discernible.
Left: Duisburg Zoological Gardens
Right: West Berlin Zoological Gardens

Overleaf:

156 Black Gibbon *(Hylobates concolor)*.
When gibbons are thirsty they dip the hairy
backs of their hands into the water and then
suck up the moisture from the fur. The variety
with white fur on the sides of the head is
classified by some authors as a subspecies.
(Liberec Zoological Gardens)

157 White-handed Gibbon or **Lar** *(Hylobates lar)*.
The lar is the most common gibbon species in zoos.
(Liberec Zoological Gardens)

Agile gibbons are found on the southern part of the Malay Peninsula and on Sumatera. They are very popular with the local people who call them "unkapati," rearing young gibbons in the family circle like their own children. The adolescent animals then often remain in the home or the immediate vicinity of the settlement for a long time. It is astonishing how they manage to adapt themselves to human fare. Agile gibbons are not very common in zoological gardens. In 1944 the Washington National Zoo possessed a hybrid resulting from the mating of an agile gibbon and a cap gibbon. It later gave birth to offspring.

According to Sanderson there is a respective subspecies for the Malay Peninsula and for Sumatera.

Silvery or **Gray Gibbon** (*Hylobates moloch*). There is no information about varieties that differ in color. Its coat is usually a uniform medium, pale to silvery gray color. On the forehead over the black face is a fairly clearly defined narrow whitish stripe. On each side of the brows there is a conspicuous tuft; the face is framed by a ruff of fur. Silvery gibbons occur on Kalimantan, Palawan and Java where the herds have become much depleted as a result of spreading human settlements. Its habits are almost identical to those of the lar and agile gibbon which are its close relations. The silvery gibbons, too, are often reared by the local people in their homes and then frequently remain in the gardens and villages. These pets are exceptionally affectionate and seek contact with the human beings who look after them. They allow themselves to be carried about everywhere and often accompany their human friends on travels and sea voyages. In zoos the silvery gibbon is less common than the lar and black gibbon. It has, however, been bred in captivity and in 1965 the Berlin Zoo successfully crossed a lar and silvery gibbon.

Views differ about the identification of subspecies. Up to eight subspecies are thought to have evolved on the various islands. This classification difficulty is partly a result of the acclimatization of gibbons brought from other regions. A subspecies is, for instance, *Hylobates moloch albibarbis*.

158 Silvery Gibbon (*Hylobates moloch*). This species has often been acclaimed for its good behavior in the home and general teachability. It has frequently been bred in captivity. (West Berlin Zoological Gardens)

Great Apes (family Pongidae)

Although the great apes have manifold features that are reflected in their different habits and their disposition, it is worth noting what they have in common. Apes are not two-legged animals but they have a direct tendency to walk erect. Their arms, which are longer than the legs, are still used in the first place as climbing, swinging, and locomotive organs. When in motion they do not use the entire surface of their hands but just support the body with the back of the hands or, more accurately, with the knuckles; the main weight is carried by the legs. Anatomical features, such as the structure of the pelvis, the shape of the backbone, parts of the skeleton as well as the structure and development of the musculature enable them to stand erect for short periods but do not equip them for permanent bipedalism. Whereas the heavy apes mainly swing beneath the branches as a means of locomotion, the monkeys run on all fours along stout branches and twigs as well as on the ground. Like the thumbs, the big toes of all apes are opposable. So there are significant differences, especially in the manner of locomotion. The chimpanzees in particular have a very precise grip, being able to turn the thumb and press it against the side or tip of the index finger (Napier 1961). In the course of their evolutionary history the apes' hand has become far more skillful and manipulative and this of course called for new physical properties. All apes have powerful canine teeth; the male's in particular are formidable and effective weapons. The ape's gestation period is between eight and nine months, and childhood and adolescence up to sexual maturity last seven to eight and a half years.

The physical and mental development is hence much quicker than that of humans. Monkeys require only four to five years for the same stages. As a rule births take place at intervals of about three to four years. The growth of the more highly developed brain requires a longer period. The power of visual communication is also much improved. Feelings, moods, intentions, the desire for contact, aggressivity, submissiveness, fear, courage, and so on are expressed in manifold and more subtle gestural language, in body postures, hand and arm movements, dancing, and other forms. Mutual grooming is more than a hygienic device—it is the "social cement" of the primates right up to the chimpanzees.

The structural evolution leading to a brain that, although it is not relatively large is definitely bigger than that of the monkeys and gibbons, produced a new and very significant factor. Experiments, with chimpanzees in particular, have challenged the previously assumed limits of their intelligence and shown that they possess astounding associative powers. This increased brain power has naturally had an effect on the apes' habits; it reveals itself, for instance, in the use and manufacture of tools. Man is not the direct descendant of today's apes; but he shares with them a common ancestor, an ape-like creature that lived from 15 to 12 million years ago, and from which the various types of existent apes also evolved.

Orang Utans (genus *Pongo*)

Orang Utan *(Pongo pygmaeus)*. Many legends tell of animals that impressed or seemed strange to human beings. In the mid-18th century rumors about strange man-like creatures in the Southeast Asian jungles began to circulate in the civilized world. The coastal inhabitants of Sumatera and Borneo talked about these creatures that they called "orang utan" (forest man). The tribes living there whose level of civilization is still very low are called "orang kubu" by the Malays, who at that time apparently made no great distinction between the arboreal-living orangs and the jungle people who lived a partly nomad life on the ground. It was believed that the orang utans were men who had abandoned speech or refused to talk in order not to have to work, and who for this reason had fled far into the jungle. The name has remained in zoology and is found in many languages.

Of the four species of apes the orang utan has the oldest lineage. Its evolution away from the common line of descent probably took place in about the latter half of the Oligocene period (mid-Tertiary), that is about 30 to 25 million years ago, whereas this process, according to present-day knowledge, began for the gorilla and chimpanzee only about 25 to 15 million years ago. The orang is the only ape found in Asia. As fossil remains indicate, its genus also inhabited the Southeast Asian mainland of Indochina, South China and occurred as far as the Indus during the Pleistocene period, surviving in some places even after the Ice Age, about one or two million years ago. Today the orang utan is only found in some parts of Sumatera and Kalimantan. Estimates about their number differ. But the maximum figure does not exceed 2,500 and is possibly much lower.

A male orang utan may stand up to 1.60 meters tall, the female has a maximum height of only 1.30 meters. The build also differs considerably according to sex. Their weight ranges between about 40 kilos for a female to 100 kilos for a male. In captivity the latter tend to get fat because of lack of exercise or too nourishing food so that they sometimes weigh a good deal more. The dusky to chestnut-brown long-haired and somewhat shaggy coat makes the body look bigger than it is. "Buschi II", an orang in Dresden Zoo, had hairs on his back measuring up to 1.20 meters. When living in the wild an orang utan's fur would scarcely grow to such length since it would be rubbed off while clambering about the branches. The fur on the lower and upper arm is pointed toward the elbow in adaptation to the angle of the arm while climbing and resting in the trees so that rainwater can run off in both directions. The arms that are used

to swing beneath the branches are very long and powerful and have a stretch of up to about 2.40 meters. The hands, which serve as hooks to latch on to the branches, have long, slender fingers. The comparatively small thumb has moved away from the other fingers down to the wrist with the result that the orang is not very dexterous in grasping and manipulating smaller objects and is less skilled in using its fingers than, for example, the chimpanzee. In comparison to the other apes the legs are weaker and somewhat shorter, but are adapted for an arboreal life by the ability to stick them out at right angles to the body. In a similar way to the thumb, the big toe is opposable to the other relatively long toes, so that as an accomplished clamberer the orang can use the foot as a grasping organ, holding on by it to lianas. The comparatively narrow but heavy skull is joined to the torso without any visible neck. The cranium is frequently surmounted in adult males by a prominent sagittal crest to which the jaw muscles are attached. The ridges over the close-set eyes are small in comparison to those of the chimpanzee and gorilla. The many separate muscles in the face enable an expressive play of features and, in particular, a variety of lip gestures. Orangs have the same number of teeth as humans. The larynx sac which, when inflated, has a capacity of up to 6 liters, is used by the adult male as a resonator and air reservoir for his hollow resounding call during the initial phase of which he can be heard "pumping up" his throat pouch. When they are about 12 to 14 years old the males develop broad cheek flanges which are formed by thick fat deposits under the surface of the skin, giving the face a rounded appearance. The size of these cheek bulges varies. At this time the males begin to display a more individual or solitary disposition; they are now fully developed. The face is almost hairless; only the adults have whiskers on their cheek and chin. From the top of the head their fur forms a fringe toward the forehead.

In their diet the orangs can adapt themselves to a certain extent to local conditions. They chiefly eat sweet and bitter fruits of the many fig species, bananas, nuts, seeds, buds, shoots, leaves, blossoms, bark, as well as lizards, tree frogs, young birds, eggs, insects, and other small animals. When the food supply is abundant the orangs tend to put on weight; when it is scarce they have to make long journeys through the forest in search of it.

With the appearance of cheek bulges and the growth of longer fur, particularly on the back, the male begins to utter his call. The tussle for dominance begins but rarely leads to embittered status struggles. It usually takes the form of brief encounters in which the orangs measure their strength or enforce subjection. At times a male will return to a solitary life before embarking on another attempt to assert his status in the hierarchy.

There is unfortunately still a lack of field studies dealing with the orangs so that considerable gaps exist in our knowledge about their habits in the wild. As a result of their brief mating relationships they have no settled family life. In the vast tracts of tropical forest the male finds more opportunities for diversion and development of his aptitudes than the female whose maternal duties in rearing her offspring (a baby and often an older child at intervals of about 4 years) fully absorb her energies. Orangs thus live in loosely knit groups that are scattered over a wide territory. Mating—which usually takes place in a hanging position—may be a completely harmonious affair but is sometimes tantamount to sexual assault. From about the second or third month of pregnancy the female's vulva, as in the case of the gibbons, is visibly swollen. After gestation lasting about eight months one baby—rarely two—is born. Weight at birth usually ranges between 1,700 and 1,900 grams. From the very first or second day the baby is capable of clinging to its mother's fur with its hands and prehensile feet. Quite helpless to begin with, it needs its mother's intense and loving care. The physical development of the orang baby differs little from that of a human infant. Many orang mothers—especially when handling their first-born—are extremely clumsy and rough or absolutely incompetent so that most orang babies born in zoos have to be reared by human hand. When patience is shown the orang mother not infrequently soon learns the maternal techniques and looks after her baby properly. In 1965 at the Dresden Zoo it was observed for the first time how an orang father ("Buschi II") acted as midwife. At the first sign of the impending birth and probably sexually stimulated by mucus secretion, the male practically assaulted the female ("Suma II"). With the start of the expulsion phase, as the top of the baby's head became visible, a complete change in Buschi's behavior occurred. Suma was now bent forwards in a crouching attitude and he used his big lips as a suction pad, fitting them to the baby's skull. He then gripped its chin and back of the head with his lips, pulling very carefully and taking advantage of the mother's labour pains, until the baby was almost delivered. He held the baby's body very gently in his big coarse hands taking care not to let it fall as it was born. He at once started to lick the amniotic fluid out of the newborn's coat. Shortly afterwards the new arrival uttered its first cry—a signal for Suma to pay attention to her child. Buschi laid it voluntarily in her arms and both parents started to dry their offspring's fur. It was a surprising and impressive event that approached or even revealed a display of intelligence. Suma repulsed Buschi's cautious attempts to touch the baby gently and he respected this too. The mother had taken full possession of her child. At subsequent births Buschi again proved himself a skillful midwife. Suma learned to cope better with each succeeding child, became more experienced and less anxious than she was after the first birth. The father was soon given the infant to hold.

This obstetrical aid was observed in all its phases on several occasions. W. Ullrich has given an account of it; I, too, had the good fortune to be present twice. It would, of course, be prema-

The orang utan *(Pongo pygmaeus)* is fitting two objects together with the aid of its lower lip (after Lethmate).

ture to draw general conclusions about this specific behavior since there is no record of anything similar either from other zoos or in the course of field studies. The number of orang births and successful rearing in captivity has only increased in the course of recent decades as a result of better conditions and improved experience, even though many young animals still have to be reared by hand. Some zoos have already reported births in the second generation. Nevertheless, orang utan zoo breeding successes are still comparatively rare.

Orangs sleep on trees at night, building a platform of branches and twigs for this purpose. These nests are usually built anew every day at a height of between 10 and 25 meters according to the structure of the tree cover. In zoos, too, they display great skill and intense activity in building nests on the ground; it is an instinctive habit and with growing experience they become remarkably proficient.

Until a few years ago the "phlegmatic" disposition and habits of the orangs and unsatisfactory results of experiments with them—especially in comparison with the chimpanzees' performance level in learning—caused them to be regarded as hardly suitable for the solution of complicated problems. In addition, the small numbers of these valuable and delicate animals prohibited the carrying out of lengthy experiments. But more recent systematic tests and studies of zoo orangs have produced some very instructive results. D. M. Rumbaugh, an American primatologist, was the first to doubt whether the chimpanzee was really the most "intelligent" animal amongst the great apes. J. Lethmate made a specific study of the orangs' approach to the solution of problems. Problem solution is the term used to describe a complex situation with a number of potential manipulations, the correct choice or discovery of which produces the solution (for instance in the use or making of a tool.) A test animal, for example, succeeded in mastering the comparatively complicated combination process of joining five sticks together with the aid of hollow tubes without being previously shown how to do so. The sticks were made to fit the tubes by biting off the ends, pointing and splitting them. When the tool was ready, the next step was to use it to angle a tasty morsel placed well away from the cage. The functional effect of a logical sequence of actions was evidently understood. There was also a rational reaction to a sudden change in the situation. With a margin for trial and error, the spontaneous switch to a new sequence of actions seems to indicate a degree of reasoning (W. Köhler). The approach to the solution of problems is also obviously conditioned by the individual animal's age. Whereas two or three-year-old animals required more or less longer periods to learn how to tackle a problem, the six to nine-year-old orangs found almost all the solutions at the first attempt (more experience, better manipulative skills). In Münster a female orang removed a dab of color on her face with water by studying her reflection in a mirror.

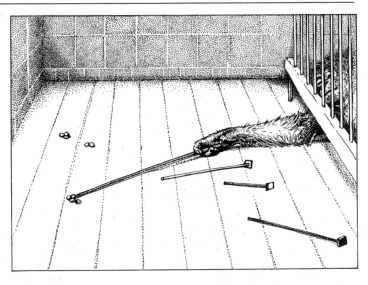

Over three intermediate stages the orang utan *(Pongo pygmaeus)* has reached the longest tool with which it can angle the bait (after Lethmate).

The reason why orangs have hardly ever been seen to use tools in their forest habitat is probably based on the previously mentioned lack of knowledge about their habits when living in the wild; in the dense tree cover it is extremely difficult to observe them. Their highly specialized arboreal life therefore does not provide an explanation; moreover, it is now known that they descend to the ground more frequently than was hitherto assumed. In any case they possess potentialities for rationally combining a sequence of actions demanding the use of tools and elementary reasoning.

Advances in the keeping of orangs were not made until about fifty years ago. Although marked progress was registered with the survival of the male orang "Peter" in Dresden Zoo from 1898 to 1907, setbacks frequently occurred. The Dresden Zoo director Professor G. Brandes who succeeded in raising an orang baby to adulthood ("Buschi") was the first to provide evidence that these sensitive charges can survive for a long time in our latitudes too. In 1927 Brandes acquired the mother orang "Suma" with her baby "Buschi" who was born during the voyage on the Red Sea. Brandes' new and biologically sound methods were strikingly successful and were soon adopted by many European and American zoos. With this hitherto unparalleled achievement Brandes became the founder of the modern method of keeping apes in captivity. Since that time zoo orangs have had a life expectancy of from 30 to 40 years. The record is held by the Philadelphia Zoo. "Guarina," a grandmother orang, died in 1976 at the age of 56. Her partner, "Guas," lived to the age of 57! This pair produced nine offspring and lived in captivity for more than 50 years. The International Studbook for orang utans recorded a total of 776 animals in more than 160 collections in 1980. 337 of these were born in the wild and 399 in captivity. About 550 orang babies have so far been born in zoos; in 1979 alone there were 55 births although not all the animals were successfully reared.

Increased breeding successes in the zoos raises the question of whether it is possible to release adult zoo orangs to their natural habitat to live in the wild. This idea is, of course, feasible. The Indonesian government placed a ban on the capture and export of orangs long ago. The International Union of Directors of Zoological Gardens (IUDZG) forbids the purchase of orangs illegally smuggled out of Malaysia. Government export licences are only granted in exceptional cases. The former methods of capture caused heavy losses. In order to capture young animals—the demand being usually limited to these—the mothers had to be shot. The young were often wounded too, or died as a result of inadequate care during transport. Another method, which was more successful, was to drive a group of orangs together, felling the surrounding trees, and catching the animals on the remaining tree with nets. But this procedure, too, entailed a certain number of losses. Even today tree-fellers sometimes catch an animal and sell it, although this

In order to reach the bait the orang utan *(Pongo pygmaeus)* has placed a crate beneath it (after Lethmate).

| howling | symbolical eating | standing erect | throwing parts of plants | drumming its chest |

Expressive gestures of a gorilla *(Gorilla gorilla)*.

is illegal. The modern abuse of attempting to transform wild animals into domestic pets still causes unscrupulous and profiteering animal trappers to hunt coveted apes for the purpose of selling them at fancy prices. The fragmented geographical structure of Indonesia and Malaysia makes it difficult to exercise control with the aim of stamping out smuggling and black market transactions. The rapid depletion of the herds alerted the international organizations for the protection of animals. In close cooperation with the World Wildlife Fund (WWF), the IUCN and regional organizations, such as the Frankfurter Zoologische Gesellschaft, several rescue centers were established with the aim of re-settling illegally purchased captive orangs in their natural habitat. In 1964 the Sepilok rescue station or "ape school" in Sabah (Kalimantan) was opened by A. G. Gorotud. But the initial move (1961) was made by Barbara Harrisson from the Sarawak Museum on Kalimantan who with a varying degree of success prepared illegally captured and subsequently confiscated orangs for their life in the wild. The successes so far achieved give rise to optimism but are still inadequate. While there are still enough illegally captured and confiscated orangs in their homeland that can be used for this purpose there is no point in sending zoo-bred orangs to the rescue stations. Until such time more orangs should be bred in zoos to meet this potential demand.

The orang utan is registered by the IUCN in the Red Data Book as a species threatened with extinction.

Two subspecies, which are not easily identifiable from a morphological point of view because even individually they differ considerably, are known:

Sumatra orang utan *(Pongo pygmaeus abeli)*, occurs in distinct areas of northern Sumatera; **Borneo orang utan** *(Pongo pygmaeus pygmaeus)*, occurs in distinct areas of Kalimantan.

Gorillas (genus *Gorilla*)

Gorilla *(Gorilla gorilla)*. Travelers to Africa had given horrifying accounts of the gorilla's ferocity long before the first living specimen arrived in Europe in 1855. Ignorant of the biological facts, people imagined it to be a "black beast, a cunning forest demon and aggressive black monster." At the beginning of the 20th century it served as a model for the horror figure of King Kong. The true nature of the gorilla was only later discovered by scientific observers and zoo biologists.

Unlike the Asian apes, gorillas and chimpanzees mainly live on the ground. Although descended from a common ancestor, they display marked differences. The two lines of development probably began to emerge already at the start of the Pleistocene period.

Gorillas have short broad hands and feet that are no longer adept at swinging beneath the branches. The longer arms mean that the animal has a sloping backline. Like the other apes, its big toe can be opposed to the other digits in the same way as the thumb; the sole of the foot lies flat on the ground. With its imposing and powerful appearance the gorilla is the largest existing primate. It stands about 1.75 meters tall; with stretched body and straightened knees it reaches a height of more than 2 meters (maximum 2.20 meters). The arms have a stretch of up to 2.75 meters; the chest measurement is about 1.75 meters. The much smaller female weighs from 70 to 140 kilos—about half that of the male who may weigh up to 300 kilos. Due to lack of exercise or a too highly concentrated diet some zoo gorillas grow fat and may weigh up to 375 kilos. The brain of a fully adult male gorilla has a weight of about 685 grams. The deepset reddish brown eyes appear comparatively small. The coat is slate gray to black. The fur on the back of a male gorilla begins

| stamping | slinging sideways | pulling up plants | drumming on the ground |

to turn silvery-gray when he is about 10 or 12 years old (silver-backs). The naked face and the hairless hands and feet are black and so is the chest of older males. The male is sexually mature at an age of 8 or 9 years, the female at 7 or 8 years of age. Pregnancy is said to last from 245 to 255 days. Since female gorillas normally have a fairly big abdomen their pregnancy is often not discovered until they give birth or, on the other hand, their bloated appearance may give rise to false hopes. Birth weight ranges between 1,800 and 2,400 grams. Twins are rarely born. Although the females possess a maternal instinct, they have to learn by experience how to look after their babies properly. This process takes place in the family group. The young mother often handles her first baby somewhat clumsily and may seem even scared of it.

When gorillas arrive at the zoos they are nearly always between 2 and 4 years old and usually grow up in groups of the same age. From whom, then, can they learn maternal care? The young mothers, and not infrequently older ones too, have no notion of how to care for their babies and sometimes even refuse to accept them. So gorilla infants often have to be reared by hand. The males, too, need to learn "sexual technique" as well as courtship and contact behavior from the adult gorillas. The males have a very small penis and it is hence hard to determine when definite mating takes place. So it would appear hardly surprising that the first zoo gorilla breeding success occured much later than in the case of other apes—in 1956 in the Columbus Zoo (USA). In Basle the first gorilla birth in Europe was registered in 1959; in the meantime the second zoo generation has become reproductive. The 11th baby since 1965 was born in Frankfurt on Main in 1980, this number including twin girls in 1967. The Basle Zoo was the first to record the natural rearing of a gorilla in 1961. Some US zoos, too, have currently reported breeding successes. Nonetheless the birth of gorillas in captivity is still one of the most uncommon ape breeding successes.

In order to induce sexually inexperienced pairs to reproduce—most zoos cannot afford to keep large groups for breeding purposes—an "educational" film was made some years ago in Basle in which mating behavior in all its phases was shown. Since apes are receptive to visual impressions, react to them and often imitate what they have seen, sexually mature gorillas were shown the film in the hope that it would stimulate them to do likewise. It is, of course, important to bear in mind that gorillas, like many other animals, chose individual mates. The results achieved so far cannot be regarded as promising unless certain reservations are made.

In this connection it should be noted that it is hard to determine the sex of young and adolescent gorillas, since there are no primary or secondary characteristics that enable reliable identification. Where the chimpanzee and orang utan bear an unmistakable indication of sex, the gorilla has a bare patch of skin with a short peg-shaped structure that could be either a penis or a clitoris. When doubts were expressed about the sex of the two mountain gorillas in Cologne Zoo, Dr. M. Gorgas applied the hair follicle test, also known as the "Olympia test." Under a microscope the chromosomes in the cells of a hair follicle are easily recognizable. Male and female cells being furnished differently with X and Y chromosomes, this test ensures accurate sex determination. For several years, "Achille," a Basle Zoo gorilla, was considered to be a male. During a stomach operation to remove a ball-point pen he had swallowed, the surgeons found ovaries: in the meantime "Achilla" has given birth to several babies. The first gorilla to result from artificial insemination was born in Memphis Zoo (USA) in 1980.

Today lowland gorillas are distributed over parts of Cameroun, Gabon, Congo, the southwest of the Central African Republic and the northwest of Zaire. The mountain gorilla's habitat is far away, in disjunct areas of the African interior—in the western territories of Uganda, Rwanda-Burundi and on the volcanic massifs marking the borderland of Zaire.

Gorillas are typical forest-dwellers. Although adept climbers, they are more frequently found on the ground than the chimpanzees. Like the other apes, gorillas cannot swim. It is extremely difficult to study in detail their habits and daily routine over a longer period. Dense thickets often hinder observation. Lowland gorillas are only found in forest belts with a damp, hot climate, and they avoid larger clearings as well as fringe areas as far as possible, although they are not averse to sampling the products of nearby plantations. Nowadays more is known about the habits of the mountain gorillas than those of the lowland gorillas.

First discovered by the German explorer von Beringe in 1902 and originally classified as a separate species by the Berlin zoologist Matschie in 1903, the mountain gorillas live at altitudes ranging from 2,300 to 4,000 meters in central equatorial Africa. In 1925 the Albert National Park covering 315,000 hectares was established near Lake Edward principally for their protection. In this highland zone the climate is predominantly bleak. Various types of vegetation cover the ground according to altitude: highland rain forest, bamboo zone, scrub, arboreal heath; trunks and branches are often covered with moss and lichen.

The gorillas are found up to the tree line and are adapted to the rigours of the climate by having thick long coats. The face, too, is more hairy, and the animals' coats are a deeper black than that of the lowland dweller. The adult male mountain gorillas, too, have silvery-gray backs. Their entirely vegetarian diet is mainly composed of shoots, young leaves and other foliage, bark, roots as well as various kinds of fruit and seeds. Large quantities of this not very nutritious food with its high cellulose content are naturally required so that the animals spend a great deal of their time feeding. This fact must be taken into account by the zoos, although in captivity they often eat food of animal origin (eggs, white meat, cottage cheese, etc.). Due to the high water content of their vegetarian diet they do not require much drinking water.

George B. Schaller spent 26 months among mountain gorillas, observing them at close quarters. Ultimately six gorilla groups had become completely accustomed to his presence. Although generally shy, these gorillas seemed to enjoy being near human settlements, roads, and plantations. Subsequent to forest clearance the new tree cover and dense undergrowth, quite apart from plantations and fields, offer more and better sources of food. So it is highly probable that the population of the gorillas in the heart of the forests is less dense than in the proximity of settlements and cultivated areas. The size of the mountain gorilla groups ranges from 3 to 30 individuals. Usually they number between 12 and 20, as, for instance, in the Kabara area. A group may be composed of one and sometimes two silver-backs—one of whom naturally occupies the dominant position—, two young adult males, six or seven females, three adolescents and perhaps five children. If they have had little experience of human beings, their curiosity is sometimes more powerful than the urge to flee. The highest-ranking gorilla, a strong male, openly stares at a stranger, even approaching to a distance of about ten meters if he feels there is no imminent danger, and then usually roars. This is the signal for the females and young to rally round their leader. For a time the group's attention is riveted on the stranger; if nothing untoward happens, the tension relaxes and the animals return to what they were doing before they were disturbed. But if the silver-back scents danger, he reacts with formidable piercing calls, stands erect, beating a resounding tattoo with his compact hands on his bare chest, tossing up pieces of wood or twigs in between. With this display of strength or intimidating behavior he attempts to compel the intruder to withdraw voluntarily. If the latter does not retreat, the gorilla may launch sham attacks. He charges toward the human being, but nearly always stops a few meters before reaching him, or dashes past him. This is probably the reason for the shooting of gorillas in past years as an act of "self-defence" motivated by fear. If people flee from these sham attacks this may incite the gorilla to follow them and bite them in the legs or rear. But since he only snaps and lets go immediately—his jaws with the powerful canine teeth are intimidating—it is usually possible to drive him off, or he withdraws of his own accord. Whatever happens, one should avoid looking him straight in the eyes for this denotes a challenge, even when he voluntarily turns away. When he is mistrustful or intentionally harassed, he glares at his opponent in a similar manner—a habit shared with human beings.

Within the group or during encounters with other groups fighting is extremely uncommon. A brief sharp look suffices, and without resorting to force the gorilla glances away and returns to whatever he was just doing. Keeping the eyes down is an appeasing gesture. When "Benno," an attractive and powerful male gorilla in Dresden Zoo, grew somewhat rough in his youthful exuberance and unintentionally hurt his female keeper, she reacted in a manner characteristic of the species by bit-

159 Sumatra Orang Utan (*Pongo pygmaeus abeli*). "Djaka" is not protesting her inability to pay a bill but displaying her long hands which are well adapted to swinging by her arms. The comparatively small thumbs are clearly visible. (Dresden Zoological Gardens)

160 Sumatra Orang Utan *(Pongo pygmaeus abeli)*.
The male is heavily bearded on the chin and
upper lip and is furnished with a voluminous
larnyx sac. (Dresden Zoological Gardens)

161/162 Orang Utan *(Pongo pygmaeus)*.
The strong set of teeth includes long and
powerful canines. The skin of the cheek swellings has
a finely folded structure.
(Frankfurt on Main Zoological Gardens)

Overleaf:

163 Borneo Orang Utan *(Pongo pygmaeus
pygmaeus)*.
The male orang's typical cheek swellings differ
individually in size. The face is almost naked.
Only adult animals have bewhiskered cheeks and
chins. From the top of the head the fur usually falls
toward the forehead in a fringe. (Cologne
Zoological Gardens)

164 Sumatra Orang Utans *(Pongo pygmaeus abeli)*.
The orang utan's childhood lasts about 3 years.
Adolescence ends when it is 7 years old.
The mother rears her young with loving care.
(Dresden Zoological Gardens)

165 Sumatra Orang Utans (*Pongo pygmaeus abeli*).
"Uschi" allows keepers she knows to hold her baby
briefly. In this way it becomes accustomed
to human beings. (Dresden Zoological Gardens)

166/167 Orang Utans (*Pongo pygmaeus*).
The facial features of the young animal differ
considerably from those of the adult.
Females and young males have no cheek swellings.
(West Berlin Zoological Gardens)

168 Orang Utan (*Pongo pygmaeus*).
On trees where the orang utan can move freely
it exploits its locomotive and climbing techniques
to the full. (West Berlin Zoological Gardens)

Previous pages:

169 Sumatra Orang Utan *(Pongo pygmaeus abeli)*.
When erect the orang cannot walk in a human
manner; it moves stiffly using its hips because the
gluteal muscles are insufficiently developed and
the feet as grasping organs are not adapted to
rolling movements. (Dresden Zoological Gardens)

170 Lowland Gorilla *(Gorilla gorilla gorilla)*.
The African gorilla's physiognomy is very
different to that of the Asian orang utan,
especially with regard to the shape of the nose.
(Duisburg Zoological Gardens)

171–174 Lowland Gorillas *(Gorilla gorilla gorilla)*.
The head of adult male gorillas appears very
massive on account of the domed bulge over the
sagittal crest, to which the large temporal muscles are
attached, and the powerful neck and jaw musculature.
The animal's imposing appearance is enhanced by
the jutting ridges over the eyes and the large nostril
swellings. (Regent's Park London)
Opposite page:
When standing and moving on all fours the gorillas,
like the chimpanzees, support themselves with their
knuckles which have developed horny callouses.
(Frankfurt on Main Zoological Gardens)

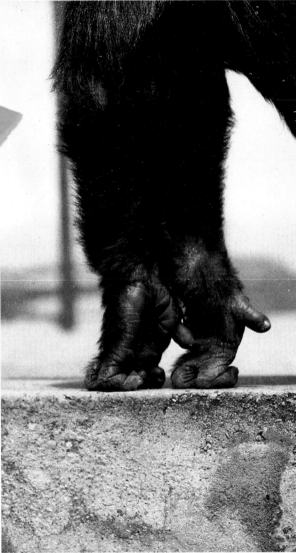

Overleaf:

175 Mountain Gorilla (*Gorilla gorilla beringei*).
A thick coat protects this animal from the
rigors of the cool and damp highland climate.
Its face, too, is more hairy than that of the
lowland form. (Cologne Zoological Gardens)

176 Lowland Gorilla (*Gorilla gorilla gorilla*).
Young animals already have the typical bulbous
nose. Reared by hand, they become very devoted to
the person who takes care of them.
(Carl Hagenbeck Zoo)

177 Lowland Gorillas *(Gorilla gorilla gorilla)*.
Every gorilla has individual features so that
it is hardly possible to confuse them.
(Frankfurt on Main Zoological Gardens)

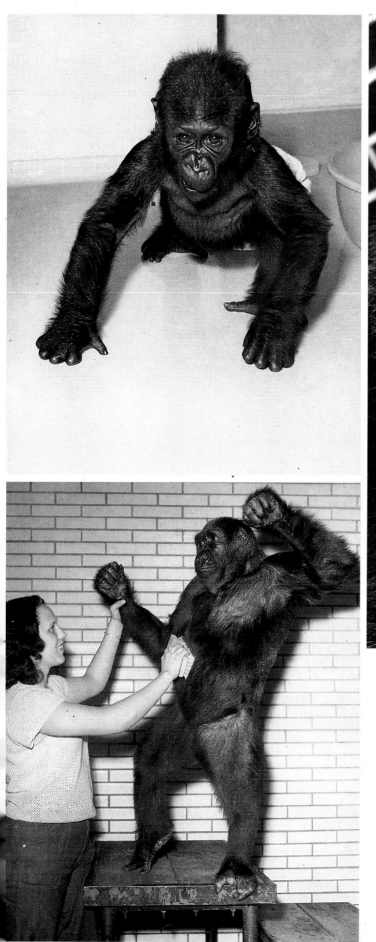

178/179 Lowland Gorillas (*Gorilla gorilla gorilla*).
Above left: When reared by hand—as frequently occurs—gorilla children provide interesting opportunities for comparison with human infants of approximately the same age.
(Frankfurt on Main Zoological Gardens)
Below left: Even though the brush is hard, this female gorilla obviously enjoys the grooming procedure. (Duisburg Zoological Gardens)

180 Mountain Gorilla (*Gorilla gorilla beringei*).
The female too has a prolific growth of facial hair.
(Cologne Zoological Gardens)

181 Bonobos or **Dwarf Chimpanzees** *(Pan paniscus)*.
Bonobo mother carefully tending her baby.
(Frankfurt on Main Zoological Gardens)

182–184 Chimpanzees *(Pan troglodytes)*.
Left: An erect posture always indicates agitation.
"Hoo-hoo" sounds uttered with pouted lips may,
according to mood, develop into loud cries
and screams. (West Berlin Zoological Gardens)
Above: "Jacky," a powerful male chimpanzee,
commands respect with unmistakable threatening
gestures, stamping, loud noises, and the hurling
of sticks and twigs. (Dresden Zoological Gardens)

185–187 Chimpanzees (*Pan troglodytes*).
Chimpanzees express their feelings with their
extremely mobile features as well as with
specific sounds and gestures. The motivations
for the respective expression of feelings
are not always identical with those producing
involuntary reactions on the part of human
beings, so that misinterpretations may occur.
Above left: Astonished, waiting to see what
happens. (Duisburg Zoological Gardens)
Above right: Irritated, partly threatening.
(Hanover Zoological Gardens)
Below right: Attentive. (Duisburg Zoological
Gardens)

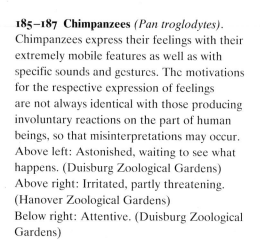

188/189 Chimpanzees (*Pan troglodytes*).
"Pouting" as a sign of pleasurable excitement
and welcoming. Conflicting feelings—joy
and fear. (Duisburg Zoological Gardens)

190–192 Chimpanzees (*Pan troglodytes*).
Young chimpanzees are always in the mood
for romping and need playmates—
their peers or the zoo-keeper as a substitute.
(Colmårdens Zoological Gardens)

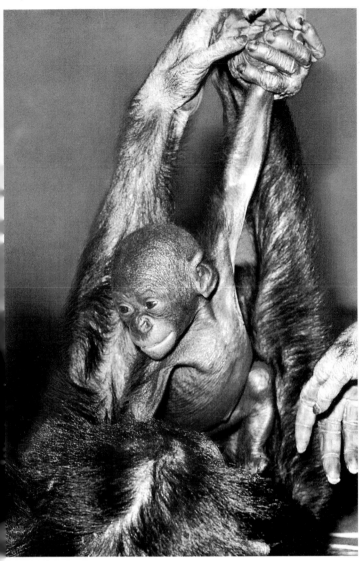

193–195 Bonobos or **Dwarf Chimpanzees**
(Pan paniscus).
Like gorilla mothers, the bonobos do gymnastics
with their babies. The bonobo baby closely
resembles a human infant. (Frankfurt on Main
Zoological Gardens)

Overleaf:

196/197 Chimpanzees *(Pan troglodytes).*
Left-hand page: Angry, stubbornly screaming.
(Brno Zoological Gardens)
Right-hand page: Screaming with fear; the
younger animal clings anxiously to the older one.
(Brno Zoological Gardens)

ing his finger. Benno was so startled and dumbfounded at his keeper's unsuspected powers of self-defence that he henceforth refrained from taking liberties with her. The gorilla's "table manners," as observed in zoos, are also very different to those of the chimpanzee, for instance. The gorilla sits in a well-behaved manner in front of his fodder portion, first using his fingers to extract the best morsels, putting them leisurely in his mouth, biting, chewing and swallowing each well-measured piece before starting with obvious relish on the next.

A very significant result of these field studies was the emergence of a completely new image of the gorilla's disposition. Despite his imposing, physically powerful and intimidating appearance, the gorilla is of a placid and gentle disposition. Through his intensive studies Schaller gained many new perceptions and recorded additional facts. Gorillas spend nearly all the daytime on the ground. They are only to be found during the night-time on the lower branches of the trees, where they build a platform usually of bent twigs every evening. The adult males stay on the ground, building their nests beneath the trees where the herd sleeps. In zoos, too, they exercise their instinctive talent for nest-building even though it is hardly ever necessary here.

In contrast to many other primate species, the gorilla's sexual behavior is of minor significance. Ritualized behavior in which the original function has acquired a new significance—such as the display of genitals, the mounting of another animal with various social motivations—has hardly ever been observed. The normally comparatively quiet black animals, however, become more vocally active when mating. The baby gorilla develops comparatively quickly. "Max," the first gorilla born in Frankfurt on Main Zoo (1965), weighed eight times as much as at birth when he was a year old; at five months he could run on all fours. The mother teaches her child and also introduces it to society. She animates her acquaintances to admire her child and to make it welcome. She clearly demonstrates her maternal tenderness and her pride in the little rascal. This is not an anthropocentric interpretation of highly developed animals' human-like behavior. They very likely do possess emotions, as H. Hediger has observed, although these are not moulded by highly developed social influences and ethical patterns as in the case of mankind.

When living in the wild female gorillas give birth at intervals of about four years. If a zoo gorilla baby is artificially reared the mother may have another baby after 18 months or two years. Young gorillas enjoy round dancing, playing tag, climbing, swinging, sliding, turning head over heels or crowning themselves with twigs. The youngsters' games are extremely important for the emergence and development of social contact be-

havior and for their later community life in the group. Only the family unit can produce a "proper" gorilla who behaves normally toward his partners. This ability is obviously also an important factor in establishing social ties, in courtship and successful mating. Segregated apes, particularly those that have been isolated from their kin since childhood, show signs of disturbed behavior, only respond to individual human beings or become neurotic. Boredom often leads to singular habits and abnormal self-absorption.

It is not fortuitous that the first zoo birth of a gorilla only occurred in 1956. Usually a zoo keeps just a pair of animals. The acquisition of several gorillas, let alone a complete group for breeding purposes, is very expensive and, as a result of the international protection regulations, only possible in exceptional circumstances nowadays. According to the most recent investigations into the reasons for the low birth rate of zoo apes carried out by Yerkes Primate Center in Atlanta (USA) in 1980, confined accommodation for these large primates is partly responsible for this state of affairs. Due to the increased population or group pressure lower-ranking animals are subjected to permanent stress and become sterile. Partitioning experiments, which enable the normally segregated females to follow their own instincts when they are sexually receptive and approach the male, have proved partially successful. Higher-ranking females, too, are said to give birth more frequently than those of lower rank. Male gorillas remain the chief problem. For instance, 62 male gorillas in a total of 48 American zoos only sired 13 offspring in 1978 and 1979.

Gorillas communicate with each other by means of 22 different sounds. In San Francisco Zoo the talented gorilla girl "Koko" had learnt 645 different signs of the American deaf and dumb language by the time she was six and a half years old. She uses 375 of these words regularly as her normal vocabulary. Under the guidance of her keeper and teacher Penny Patterson (1978) "Koko" uses both the expressions she has learnt and also new sign-words she has made up by herself, thus testifying to her gift of combination as well as to an ability to generalize. As in all similar experiments, "Koko" labours under a vast handicap in her efforts to communicate: the structure of her vocal organs and the lack of a speech control area in the brain inhibit the articulation of comprehensible human sounds. Unlike chimpanzees, gorillas living in the wild have so far never been seen to use sticks and stones as tools.

In 1980, "Massa" in Philadelphia, the oldest zoo gorilla so far, was 50 years old. Life expectancy in the wild is probably much shorter. The leopard is the gorilla's only enemy in the animal kingdom; it is said to steal small gorillas in the night, but there are very few substantiated instances of this.

This figures given for the number of lowland gorillas in the wild naturally vary a great deal. The actual number probably does not exceed 15,000 to 18,000 whose existence is regarded as threatened. The number of mountain gorillas lies between a mere 500 and 3,000 (?). The latter belong to the most endangered varieties. Forest clearance is the main reason for the ousting of gorillas from their habitat. Protective zones, particularly for the mountain gorillas, are the Albert National Park, the Kahuzi Biega National Park and the Kigesi Reservation.

In zoological gardens gorillas are still counted among the most valuable and rare animals. In 1980 488 western lowland gorillas in 113 zoos and 12 eastern lowland gorillas in 4 zoos were registered (census of rare animals in captivity). During recent years the birth rate of zoo gorillas has shown a rising trend.

For many years an albino gorilla named "Copito de nieve" (snowflake) has lived in the Barcelona Zoo. Captured in the mid-nineteen-sixties at the age of about two years in the tropical forests of Río Muni (equatorial Guinea), the albino grew up normally in the zoo. In 1973 he became a father. His offspring has the usual dark-colored coat (A.J. Riopelle 1970). As already indicated, three gorilla subspecies are today classified:

Western lowland gorilla *(Gorilla gorilla gorilla)*, western equatorial Africa, locally in Cameroun, Gabon, Congo, the Central African Republic (SW) and Zaire (NW); **eastern lowland gorilla** *(Gorilla gorilla graueri)*, which is dissimilar to the other two subspecies in the structure of the skull and body. Found to the west of the Virunga Mountains from the northwestern part of Lake Tanganyika via the forested highland region in the eastern part of Zaire, the Kayonza forest in southwestern Uganda to disjunct areas to the west of the Virunga volcanoes Nyiragonga and Nyianulagira.

Mountain gorilla *(Gorilla gorilla beringei)*, more thickly coated. Virunga Mountains, southwestern Uganda, Mt. Kahuzi region, the high mountain region to the north and northeast of Lake Kiwu.

Chimpanzees (genus *Pan*)

Chimpanzee *(Pan troglodytes)*. Chimpanzees are the most human-like apes. They have not moved so far away from the original common ancestors of apes and men as the other species.

The first chimpanzee, and indeed the first ape to reach Europe, was brought by the Dutch seafarer Nicolas Tulp in 1641. Up to the early 19th century people made no clear distinction between chimpanzees and gorillas. To begin with, there was no knowledge about the existence of three species of these great apes in Africa. Since then and up to the past two decades the image of the chimpanzee has undergone a profound change. Subsequent to the experimental research, intelligence tests and studies of its social behavior carried out in research centers and zoological gardens—for instance at the Koehler Ape Center on Tenerife (1912–1920), Yerkes Regional Primate Center in Atlanta (USA) or at the research center founded by J.P.Pav-

With gymnastic exercises the gorilla mother encourages her child to participate actively (after Lang, Schenkel, and Siegrist).

lov in Sukhumi (USSR)—observations and investigations, particularly during the nineteen-sixties, of chimpanzees living in the wild have brought to light sensational new pieces of knowledge. The names of Jane van Lawick-Goodall, the British zoologist, the Dutchman A. Kortlandt who conducted experiments with wild-life chimpanzees with respect to their use of tools, and of V. and F. Reynolds with their studies of chimpanzees in the Budongo Forest of Uganda (1963/64) deserve special mention in this connection.

Some zoological data to start off with. Chimpanzees have a body length of from 70 to 95 cm. A male may stand up to 170 cm tall, a female up to about 140 cm, although among the latter there are sometimes animals of a very powerful build. The strong arms are longer than the sturdy muscular legs. With opposable big toes the foot is an efficient grasping organ enabling the animals to climb adeptly and swiftly. Like the gorillas, they move on all fours on the ground, their hands knuckle-down. They sometimes go erect for short distances, but without straightening their knees; they appear smaller than they really are. The blackish brown or mostly black fur is thick and stands on end when the animal is agitated. The face, surface of the hands and feet as well as the anal and genital regions are naked. In infancy and childhood the naked skin is almost always light-colored and turns darker or almost black or remains patchy later on. The chimpanzee has very powerful jaws—with 32 teeth like human beings—and can inflict deep wounds with them. Strong males may weigh up to 80 kilos; females average 53 kilos, in exceptional cases a maximum of 60 kilos. The female chimpanzee has bigger swellings on her hindquarters than any other primate. These genital and anal swellings inflate into naked pink fleshy cushions the size of a baby's head when she is sexually receptive. Similar to human beings, the sexual cycle, in which estrus and menstruation alternate, lasts 28 days. The female attains sexual maturity at 7 or 8 years of age as a rule, the male at 8 or 9 years, and the first birth usually occurs a year later. The female remains fertile until she is about 40, as established by the case of "Patty," a female chimpanzee in Yerkes Laboratory at Orange Park in Florida. Chimpanzees that bring up their own children normally give birth at intervals of 3 or 4 years. Like other apes, the females do not conceive within this period. Only when they lose their offspring during the first two years of its life do they again give birth at an earlier stage. Pregnancy lasts between 240 and 255 days. Other females in the group act as midwives as the situation demands or the mother herself copes with the birth according to whether she is sufficiently experienced. The vagina is stretched open with the fingers and the baby carefully pulled out. The average weight at birth ranges from 1,800 to 2,200 grams. The birth of twins has also been registered in a few cases. The tie between the young and their mothers remains until the offspring are from 4 to 6 years old.

The provision of proper conditions and care including regular veterinary supervision—as well as consultations with medical practitioners in the case of apes—have substantially extended the life expectancy of chimpanzees and all other apes in captivity. Jones (1968) reported the case of a 46-year-old male chimpanzee in Chicago who showed barely any signs of senility. In the wild, on the other hand, average life expectancy is curtailed by illness, injuries, pursuit and other factors. Senile chimpanzees whose age could, of course, only be roughly estimated, have very rarely been observed in their natural habitat.

The blood provides evidence of one of the many anatomical and physiological similarities and affinities with human beings. Its proteins establish a major criterion for the degree of kinship between similar living creatures. In this case there is an actual "consanguinity." The serum protein in the blood of the gorilla and chimpanzee is far more similar to that of human beings than that of the orang utan, as J. Schmitt established in Frankfurt on Main. There are only minor differences between the blood of chimpanzees and bonobos. But the serum protein in bonobo blood is more akin to human blood than that of the chimpanzee. In the same way the convolutions and finer structure of the chimpanzee brain are similar to that of human beings. The mass of the brain of a human male averages between 1,350 and 1,500 grams, whereas that of a chimpanzee is normally only about 400 grams. Nonetheless, aside from human beings, the apes have the highest potential qualities.

The distribution of the chimpanzees is limited to equatorial Africa; in the west from Guinea to the Ivory Coast and in a disjunct area extending from Cameroun, Gabon via northern Zaire to Uganda, Rwanda, Burundi and Lake Tanganyika. They have been exterminated in various regions during recent years. Nowadays zoologists make a distinction between two regional chimpanzee groups with a very different habitat; populations in fairly compact forest belts and those found in the less dense tree cover of forest edge areas as well as in the open country of the savannahs. In adaptation to the various ecological conditions their habits and specific behavioral patterns have undergone changes which also reveal interesting evolutionary trends. Kortlandt (1967) placed a dummy leopard on a path used by chimpanzees. When the apes arrived on the scene their fur bristled, they roared, threatened and beat the dummy with sticks, raining heavy blows on its back that could have inflicted severe injuries. Forest chimpanzees would probably have sought safety in the trees, whereas the populations of the open plains cope with such situations by wielding sticks and hurling stones as weapons of attack, intimidation or defence. Chimpanzees possess a broader spectrum of endowments and aptitudes than are normally brought into play. The open plains with their more multifarious changes and hazards call for a higher development of the senses and mental capacities. Dense forest, on the other hand, guarantees more peace and protection but calls for less resourcefulness and action to ensure survival.

The outstretched hand of the higher-ranking male chimpanzee
has a pacifying effect on the female ...

When living in the wild, chimpanzees learn by experience to use and make certain tools. Stalks or thin twigs are trimmed with the teeth or bent to the right shape and poked into the holes of termite hills, and are used in the same way to extract honey. Chimpanzees chew leaves to a porous mass for use as drinking sponges. Sticks and stones are intentionally manipulated for throwing, hitting and beating purposes; children sometimes play with them. Sticks are also used as smelling aids; to test, for instance, when termite holes are occupied. Leaves are often used to wipe excrement and dirt from the body. Inanimate objects may be used as weapons.

Chimpanzees build platforms in the trees to sleep on, rarely on the ground. As they grow older their ability to build platforms improves, along with other manipulative skills. Young animals are not left alone on the platform. Mainly vegetarian in their diet, chimpanzees eat fruit, leaves, blossoms, young bark, pithy stalks, etc. Their insect fare consists chiefly of ants and termites which, as already mentioned, grip the twigs or stems that are poked into their holes by the chimpanzees who suck off the termites by drawing the stem through their teeth. Chimpanzees only take food they are familiar with. For instance, oil palm fruit, eaten with relish by chimpanzees in West Africa, is rejected in other regions; some eat eggs, others spurn them. Chimpanzees on the open plains even capture and eat small mammals like bush pigs, young baboons, and newborn antelopes. A red colobus monkey was once found among their prey (Kawabe 1966). They divided up the flesh without quarreling and chewed it slowly. After each bite a bunch of leaves was eaten. Hunting, which only occasionally occurs, is generally a community affair. At the kill the outstandingly strong males act like big cats, tearing the prey apart or thumping it hard on the ground. Successful hunters sometimes claim priority in the division of flesh. Even dominant animals, however, do not snatch the food but sit with outstretched hand, sticking out their lower lips in a pleading gesture. The flesh is often divided into pieces whereby begging females usually get a smaller portion.

On the basis of their genetic endowment chimpanzees should be able to use sticks and stones as hunting tools. Kortlandt (1963) considers that earlier on apes used more tools and weapons than they do today. In his view the chimpanzees show a retrogressive trend in their manipulative and mental powers. Opinions differ as to why apes living in the wild do not manipulate more objects than they do in captivity. Human beings promote and systematically steer the chimpanzee's learning process, organize teaching programs, give guidance and methodically utilize its inquisitive habits and manipulative urge. Rewards and incentives in the form of food, etc. stimulate learning activity. Under natural conditions, the normal social contacts provide the only stimulus for simple learning processes, imitation, the passing on of experience, faster reactions in the face of danger.

The behavioral studies of chimpanzees undertaken by Jane van Lawick-Goodall have produced valuable results. With the exception of a few brief intervals she worked for ten years in the Gombe River Reservation near the northeastern shores of Lake Tanganyika in Tanzania. As from 1960 she studied and observed a group of about fifty chimpanzees whose initial suspicion was gradually superseded by an astounding degree

... who humbly kisses it.

of intimacy; they ultimately accepted her almost as one of their own.

The chimpanzee community observed by Jane van Lawick-Goodall (1971) was not a unified group but was divided up into several clans numbering from about three to six animals. At places where they congregated for feeding and other purposes, however, groups numbering twenty and more animals were observed. The dominant role there was played by a powerful aggressive male, and also sometimes by a weaker animal whose wily tactics made up for his physical inferiority. When mutually acquainted chimpanzees meet while on the move they welcome each other in a surprisingly human manner. There are embraces, kiss-like touches on the mouth, outstretched hands or hand-shakes, jovial pats on the shoulder. Physical contact provides a feeling of security. On the other hand, the males sometimes behave in a threatening manner that is designed to inspire fear and may be the prelude to attacks or serious clashes. A mollifying response, or, better still, submissive gestures, such as the presentation of hindquarters, usually succeed in pacifying aggressive males. The loser either seeks refuge with his friends in the community or flees, but only withdraws from his group with reluctance. Often the loser vents his anger at his humiliation on an even weaker member of the group. It is known that chimpanzees fight over domination status in their clan, over a female or even over food. Such status rivalries often occur but rarely involve a trial of strength with severe bites and injuries. Generally speaking, unity prevails in the clan. It is a very different case when there is a clash between members of various groups. For unknown reasons the chim-

panzee community observed by Jane van Lawick-Goodall in the nineteen-sixties split into two groups between 1970 and 1972. To begin with, the two groups encountered each other on the borders of their territories without open hostility. The males contented themselves with threatening gestures. In the ensuing years, however, there were outbreaks of warlike violence. At shorter or longer intervals males as well as females were captured and killed by blows and biting. By 1977 the one group had been completely exterminated. The reason for this totally aggressive behavior was by no means plain. I. Eibl-Eibesfeldt (1975) calls this intra-species aggression a "warlike conflict."

It has repeatedly been observed (Suzuki 1971) that individual females capture, kill and eat the children of other chimpanzee mothers, usually when the latter are unaccompanied. Conflicts may also occur when isolated groups encounter each other. But usually the battle is fought in terms of noise, display, threats and sham attacks or, according to the circumstances, evasion and retreat. Individual young females, especially those that are sexually receptive, can also visit strange groups without fear of reprisals. A status hierarchy obviously exists, even though it is not always fully apparent. Chimpanzees can therefore turn very aggressive, as is indicated by the many forms of pacifying behavior, the welcoming and appeasing gestures; for instance, lower-ranking animals extend a hand when they want to appease a superior in a critical situation. If the superior touches the hand, the lower-ranking chimpanzee is relieved. Embracing, touching, grooming, crouching down, sexual presentation of the hindquarters, etc. are other gestures of appeasement or submission. Display or threatening behavior visually emphasizes or exaggerates self-esteem and helps to intimidate the opponent. The acoustic impact is heightened by drumming on hollow tree-trunks, empty receptacles and rumbling walls. Such actions are connected with the urge to attract attention which is particularly noticeable in captivity and is utilized, for instance, in the case of performing animals. "Jacki," a male chimpanzee in Dresden Zoo, used to pose at the top of the climbing frame, looking around to see if he was being admired, before jumping down with an air of bravado. If he was applauded and cheered on he repeated his performance until he was completely exhausted. His mate often had to calm him down afterward. In such exciting situations the penis is frequently erected but this is a display symbol that has no sexual connotation.

D. Morris (1962, 1971) conducted extensive drawing and painting tests with "Congo" and came to the interesting conclusion that chimpanzees do not just paint out of a random urge to amuse themselves. "Congo" carried out his tasks with intense concentration accompanied by sounds of enjoyment and also knew when his "work of art" was finished. Chimpanzees have a clearly defined and recognizable feeling for design and even an individual style; they can vary shapes, copy patterns or round them off. Favorite themes, like radial lines or elementary color compositions can also be identified. In Sweden the pictures painted by the chimpanzee "Pierre" were exhibited under the name of Pierre Brassau in 1964. They were even reviewed by art critics and some of them found purchasers. B. Rensch and J. Döhl (1969, 1975) analyzed the "paintings" of their intelligent chimpanzee "Julia" and discovered basic compositional elements in them. All these animals obviously enjoyed painting.

It is here appropriate to recall the "classic" intelligence tests carried out by Wolfgang Köhler on Tenerife during the First World War. His celebrated experiments in which chimpanzees learnt to pile boxes on top of each other or to join two sticks in order to reach a banana formed the basis of his theory of learning by stimulating creativity.

Chimpanzees use a wide variety of facial expressions to denote their moods: laughing, smiling, joy, attentiveness, irritation, annoyance, crying, sadness, fear, anger, etc. The psychologist N. Kohts mimed a broad spectrum of emotions with facial expressions in front of a young chimpanzee that reacted to them with understanding.

Various aspects of the chimpanzee's psychological performance have already been mentioned. Its genetic endowments for learning are very extensive and are stimulated and intensified by manifold environmental factors, danger, inquisitiveness, urge to play and to seek occupation, etc. Even though it differs individually, the astounding memorizing capacity enhances the ability to make use of what has been learnt and enables the animals to cope rationally with new and more complicated situations; they also become more practised in solving complex problems and are even able to make "inventions." The components of these acquired and more variable habits then carry more weight than those of the more stereotype instinctive behavior. In this connection it should be noted that isolated chimpanzees reared artificially from babyhood on have no opportunity to learn natural social behavior. They rarely become accustomed to living with their peers later on. These outsiders do not understand the characteristic contact behavior of their species, usually react incorrectly to contact gestures, thus arousing the group's aggressivity and resistance, which not infrequently leads to severe injuries caused by bites. In the large chimpanzee group at the Arnhem Zoo (Netherlands) the attempt has been made to entrust rejected newborn babies to the care of chimpanzee foster-mothers with experience of maternal techniques. The children thus grow up in the community which by setting examples for them to observe and copy plays an important role in teaching them how to behave socially.

Under laboratory conditions chimpanzees have a greater capacity for learning than in the wild. In numerous laboratory experiments psychologists have analyzed the complexity of the chimpanzees' learning process in carefully controlled and nonsocial situations and have achieved tangible results. "Julia," the previously mentioned female chimpanzee, successfully carried out complicated labyrinth tests. In this connection Rensch and Döhl (1968) noted a definite disposition to act with foresight. This chimpanzee was also able to open a larger number of boxes arranged in varying sequences by the rational selection of five different tools; she displayed an outstanding grasp of causal and logical associations. She developed her tactics visually and she drew conclusions that enabled her to open unusual boxes with miniature keys and equally tiny locks. She became astoundingly nimble with her fingers, especially when manipulating tiny screws. Locks in general seem to be very fascinating objects for chimpanzees, as some unpleasant surprises in zoos demonstrate. In Yerkes Regional Primate Center in Atlanta a group of chimpanzees learnt to use vending machines. The animals soon grasped the fact that the variously colored coins represented different values in terms of purchasing power, having noted that the insertion of a certain coin produced a larger portion of food from the machine than if another coin was used. When satiated, some of the group began to save their "money" (J. B. Wolfe). Others, who had spent all their coins, or had a bigger appetite, began to beg coins from the "savers."

"Butschi," a very teachable female chimpanzee, who had been a performing animal before she came to Dresden Zoo, was even capable of giving tuition. She taught her son "Fips" part of her repertoire, like handstands and other acrobatics, but also some bad habits such as deliberately spitting and urinating in the direction of visitors.

It is impossible to give a complete list here of all the experiments that have been conducted. In order to pursue behavioral studies that are not governed or influenced by laboratory conditions, chimpanzees are set free every year from spring to autumn on a wooded island in a lake near Pskov (northwestern USSR). Staff members of the Leningrad Institute of Physiology of the USSR's Academy of Sciences make special studies here of the apes' adaptive powers, the use of tools and the communication system. The animals can roam at will on the island and have adapted themselves to the natural environment; they build platforms for sleeping and have become acclimatized to a great extent.

A few words about the chimpanzees' "language." Their extensive and advanced communication system with the aid of sounds, gestures, facial expressions, sense of smell, touch, and so on, is well known. Each message is generally composed of motivation and emotion, signal, significance, and function (Smith 1968). That is the first level of the signal or communication system. The second level is language itself, which is restricted to human beings. But the chimpanzees possess certain elementary endowments that govern cerebral development and performance. Their potential brain-power and human-like behavior in many situations caused American neurobiologists and psychologists to conduct systematic language experiments after the Second World War. Between 1947 and 1954 C. and K. Hayes brought up "Viki," a chimpanzee girl, like their own child at home. But it cost a great deal of effort to teach "Viki" four words: mama, papa, cup (drinking) and up (expressing the desire to be picked up) which she only articulated indistinctly. Chimpanzees lack the anatomical structure of the human larynx, tongue and palate which are adapted to articulated speech. A more promising method was adopted by B. and R. Gardner from Nevada University (USA) between 1966 and 1970. They taught "Washoe," a young female chimpanzee, the American Sign Language (ASL), the deaf and dumb language, which is based on the higher mental powers of human beings and calls for conceptual thought, or at least conceptual association. First of all "Washoe" learnt proper names and the names of things in her surroundings; this was followed by verbs, pronouns and other parts of speech which specify the relationship between the terms. In four years she learnt more than 132 ASL signs which she used correctly of her own accord without having to have them repeated to her, also employing them in new contexts in an amazingly independent manner. "Washoe" thus provided clear evidence of an elementary or at least potential endowment with the ability to generalize which, together with the means of communication, is the point of departure for any kind of language. H. Terrace (1980) from the New York Columbia University came to the conclusion after careful examination of his own results that despite amazing individual performances there can be no question of a genuine language that even approximates to human speech. The Californian psychologist D. Premack concluded his research report by remarking that since man has to teach the chimpanzee to speak and not vice versa we may continue to claim uniqueness.

Although the numbers of chimpanzees do not give rise to alarm, their retrogressive trend nevertheless calls for effective protective measures. In some of their natural habitats they

The chimpanzee (Pan troglodytes)
pokes a stem into a termites' hill.
The insects grip the stem
and the chimpanzee sucks them off.

have already become extinct (Rwanda, Burundi, Benin), in other countries they are very rare (Sierra Leone, Togo, Ghana) and although their numbers are fairly stable in other regions they are only plentiful in comparatively limited areas (Guinea, Cameroun, Gabon, Congo, Zaire). The total herds probably do not exceed 50,000 animals. Like the orang utans, illegally imported, confiscated and zoo-bred chimpanzees have been returned to the wild. In 1966 B. Grzimek brought an initial group of ten adolescent chimpanzees to the island of Rubondo covering an area of 240 square kilometers in Lake Victoria. Formerly a wild-life reserve this island, along with other isles, became a National Park in 1977—the tenth in Tanzania. After initial difficulties resulting from the animals' previous contacts with men and their complicated social behavior, this release venture may now be considered a success. In the meantime the second generation of these "immigrants" has produced progeny. In Gambia, too, chimpanzees born in captivity have been prepared for the life in the wild. Under the guidance of Stella Brewer the "returnees" receive a thorough training in forest existence and acquire those habits which are not innate but essential in natural surroundings. This is the only way they can cope with and survive wild-life conditions.

Chimpanzees can adapt themselves to various habitats and have the widest distribution of all apes. In the Ruwenzori Mountains they are found at altitudes up to 3,000 meters. An unusual feature of this species is its wide range of divergent physical features (shape of body and face, size, pale or dark faced, bewhiskered, beardless, etc.) irrespective of normal sex and age distinctions. According to the superior numbers of one or the other types inhabiting the various regions some zoologists classify them into three or four very dubious subspecies:

Pan troglodytes troglodytes, Pan troglodytes verus, and *Pan troglodytes schweinfurthi.* The last named is the most common ape in zoological gardens. As a result of its comparatively good adaptability and fairly low purchase price it is kept under unsatisfactory conditions in many places where these intelligent and sensitive animals usually pine away in consequence. Despite more stringent international controls, chimpanzee babies are still sold to private purchasers.

Bonobo or **Dwarf Chimpanzee** *(Pan paniscus).* This ape was discovered in 1928. It was only during the past 25 years that more was found out about these probably most human-like apes through studying their habits in captivity. Very little is known about their life in the wild.

Dwarf chimpanzees are about one meter tall and of slender build; their weight rarely exceeds 50 kilos and the females weigh a great deal less. The forehead is higher and more domed, the face is black with light-colored chin and eyelids and the lips are reddish. The ears are smaller and the teeth are more like those of human beings than in the case of the chimpanzee.

The female's outer genitals are located more in the region of the abdomen. As a result the mating procedure is somewhat different and, according to R. Kirchshofer and Pournelle, takes place in a lying or crouching position whereby the female stimulates the male by lying flat on the ground and holding him in her arms.

Dwarf chimpanzees are found in the heart of Africa, in the impenetrable rain forests between Central Zaire (Congo) and the lower course of the Kasai river. They probably live in smaller groups of only three to five animals. At night they sleep on platforms which they usually build 20 or 30 meters up in high trees. Apart from their chiefly vegetarian diet, they also eat insects and small vertebrates.

Their calls also differ from those of their larger relations. According to H. Mitchell (1976), the dwarf chimpanzees can do more with their voices than any other chimpanzees. They have a larger repertory of sounds and sometimes laugh outright. They responded far better to Mitchell's stimuli and he found them more intelligent and much more friendly toward human beings. In disposition they are far more sensitive than their more robust relations. For many years exact investigations into the habits, blood chemistry and reproduction of dwarf chimpanzees have been carried on by G. H. Bourne at Yerkes Regional Primate Research Center in Atlanta. Because Zaire does not permit the direct export of these animals, the Yerkes Center was only able to obtain the loan of five dwarf chimpanzees from the Institute for Research of Central Africa.

Valuable observations about reproduction have already been made in the few zoos that possess these chimpanzees. The birth takes place in much the same way as among chimpanzees. Little information is available about life expectancy. In San Diego Zoo a dwarf chimpanzee lived to the age of 21, having spent 20 years in captivity. A female in Frankfort on Main was about 28 years old in 1981. In 1980 34 bonobos were registered in 8 zoos with breeding successes in five cases. The IUCN has registered the dwarf chimpanzee as an endangered species.

Supplement

Classification of Primates

	Primates	Primates
Order	Primates	Primates
Suborder	Prosimians	Prosimiae
Infraorder	Tupaiiformes	Tupaiiformes
Family	Tree shrews	Tupaiidae
Subfamily	Bushy-tailed tree shrews	Tupaiinae
Genus	Tupaias	*Tupaia*
Species	Common tree shrew	– *glis*
	Günther's tree shrew	– *minor*
	Mountain tree shrew	– *montana*
	Nicobar tree shrew	– *nicobarica*
	Javanese tree shrew	– *javanica*
Genus	Greater tree shrews	*Tana*
Species	Tana tree shrew	– *tana*
Genus	Indian tree shrews	*Anathana*
Species	Elliot's tree shrew	– *ellioti*
Genus	Philippine tree shrews	*Urogale*
Species	Everett's tree shrew	– *everetti*
Genus	Smooth-tailed tree shrews	*Dendrogale*
Species		– *murina*
		– *melanura*
Subfamily	Pen-tailed tree shrews	Ptilocercinae
Genus	Pen-tailed tree shrews	*Ptilocercus*
Species	Pen-tailed or feather-tailed tree shrew	– *lowii*
Infraorder	Lemuriformes	Lemuriformes
Family	True lemurs	Lemuridae
Subfamily	Dwarf lemurs	Cheirogaleinae
Genus	Mouse lemurs	*Microcebus*
Species	Lesser mouse lemur	– *murinus*
	Coquerel's mouse lemur	– *coquereli*
Genus	True dwarf lemurs	*Cheirogaleus*
Species	Greater dwarf lemur	– *major*
	Fat-tailed dwarf lemur	– *medius*
	Hairy-eared dwarf lemur	– *trichotis*
Genus	Phaner	*Phaner*
Species	Fork-marked dwarf lemur	– *furcifer*
Subfamily	Typical lemurs	Lemurinae
Genus	True lemurs	*Lemur*
Species	Brown lemur	– *fulvus*
	Black lemur	– *macaco*
	Mongoose lemur	– *mongoz*
	Crowned lemur	– *coronatus*
	Ring-tailed lemur	– *catta*
	Red-bellied lemur	– *rubriventer*
Genus	Gentle lemurs	*Hapalemur*
Species	Gray gentle lemur	– *griseus*
	Broad-nosed gentle lemur	– *simus*
Genus	Ruffed lemurs	*Varecia*
Species	Ruffed lemur	– *variegata*
Genus	Weasel or Sportive lemurs	*Lepilemur*
Species	Greater weasel lemur	– *mustelinus*
	Lesser weasel lemur	– *ruficaudatus*
		– *dorsalis*
		– *rufescens*
		– *leucopus*
		– *microdon*
		– *septentrionalis*
Family	Indrisoid lemurs	Indriidae
Genus	Sifakas	*Propithecus*
Species	Verreaux's sifaka	– *verreauxi*
	Diademed sifaka	– *diadema*
Genus	Woolly lemurs	*Avahi*
Species	Avahi	– *laniger*
Genus	Indris	*Indri*
Species	Indri	– *indri*
Family	Aye-ayes	Daubentoniidae
Genus	Aye-ayes	*Daubentonia*
Species	Aye-aye	– *madagascariensis*
Infraorder	Lorisform lemurs	Lorisiformes
Family	Lorises	Lorisidae
Genus	Slender lorises	*Loris*
Species	Slender loris	– *tardigradus*
Genus	Slow lorises	*Nycticebus*
Species	Slow loris	– *coucang*
	Lesser slow loris	– *pygmaeus*
Genus	Angwantibos	*Arctocebus*
Species	Angwantibo or Golden potto	– *calabarensis*
Genus	Pottos	*Perodicticus*
Species	Potto	– *potto*

Family	Galagos or Bush-babies	Galagidae
Genus	Galagos	*Galago*
Species	Senegal bush-baby	– *senegalensis*
	Great or Thick-tailed bush-baby	– *crassicaudatus*
	Allen's bush-baby	– *alleni*
	Demidov's bush-baby	– *demidovii*
	Western needle-clawed bush-baby	– *elegantulus*
	Eastern needle-clawed bush-baby	– *inustus*

Infraorder	Tarsiers	Tarsiiformes
Family	Tarsiers	Tarsiidae
Genus	Tarsiers	*Tarsius*
Species	Philippine tarsier	– *syrichta*
	Western tarsier	– *bancanus*
	Eastern tarsier	– *spectrum*

Suborder	Monkeys and Apes	Simiae

Infraorder	New World or Flat-nosed monkeys	Platyrrhina
Family	Cebidae	Cebidae
Subfamily	Night and Titi monkeys	Aotinae
Genus	Night monkeys	*Aotes*
Species	Douracouli	– *trivirgatus*
Genus	Titi monkeys	*Callicebus*
Species	Collared titi monkey	– *torquatus*
	Red titi monkey	– *cupreus*
	Orabussu titi monkey	– *moloch*
	Masked titi monkey	– *personatus*

Subfamily	Saki monkeys	Pithecinae
Genus	Saki monkeys	*Pithecia*
Species	Hairy saki monkey	– *monacha*
	White-faced saki monkey	– *pithecia*
Genus	Bearded saki monkeys	*Chiropotes*
Species	Black saki monkey	– *satanas*
	Red-backed saki monkey	– *chiropotes*
	White-nosed saki monkey	– *albinasa*
Genus	Uakaris	*Cacajao*
Species	Red uakari	– *rubicundus*
	Scarlet-faced or Bald uakari	– *calvus*
	Black-headed uakari	– *melanocephalus*
	Black uakari	– *roosevelti*

Subfamily	Capuchin monkeys	Cebinae
Genus	Squirrel monkeys	*Saimiri*
Species	Squirrel monkey	– *sciureus*
	Red-backed squirrel monkey	– *oerstedii*
	Madeira River squirrel monkey	– *madeirae*
	Black-headed squirrel monkey	– *boliviensis*
Genus	Capuchins	*Cebus*
Species	White-throated capuchin	– *capucinus*
	White-fronted capuchin	– *albifrons*
	Weeper capuchin	– *nigrivittatus*
	Apella or Fawn monkey	– *apella*

Subfamily	Howler monkeys	Alouattinae
Genus	Howler monkeys	*Alouatta*
	Guatemala howler monkey	– *villosa*
	Mantled howler monkey	– *palliata*
	Red howler monkey	– *seniculus*
	Brown howler monkey	– *fusca*
	Rufous-handed howler monkey	– *belzebul*
	Black howler monkey	– *caraya*

Subfamily	Spider monkeys	Atelinae
Genus	Woolly monkeys	*Lagothrix*
Species	Woolly monkey	– *lagothricha*
	Peruvian Mountain woolly monkey	– *flavicauda*
Genus	Woolly spider monkeys	*Brachyteles*
Species	Woolly spider monkey	– *arachnoides*
Genus	Spider monkeys	*Ateles*
Species	Central American or Geoffroy's spider monkey	– *geoffroyi*
	Brown-headed spider monkey	– *fusciceps*
	Long-haired spider monkey	– *belzebuth*
	Black spider monkey	– *paniscus*

Family	Callimiconidae	Callimiconidae
Genus	Callimico	*Callimico*
Species	Goeldi's monkey	– *goeldii*

Family	Marmosets and Tamarins	Callithricidae
Genus	Marmosets	*Callithrix*

Species	Common marmoset	– *jacchus*
	Buff-headed marmoset	– *flaviceps*
	Black-penciled marmoset	– *penicillata*
	White-eared marmoset	– *aurita*
	White-fronted marmoset	– *leucocephala*
	White-shouldered marmoset	– *humeralifer*
	White-necked marmoset	– *albicollis*
	Santarem marmoset	– *santaremensis*
	Yellow-legged marmoset	– *chrysoleucos*
	Silvery marmoset	– *argentata*
Subgenus	Pygmy marmosets	*Callithrix (Cebuella)*
Species	Pygmy marmoset	– *pygmaea*
Genus	Maned tamarins	*Leontideus*
Species	Golden lion marmoset	– *rosalia*
	Golden-headed tamarin	– *chrysomelas*
	Golden-rumped tamarin	– *chrysopygus*
Genus	True tamarins	*Saguinus*
Subgenus	Negro tamarins	*Saguinus (Saguinus)*
	Negro tamarin	– *tamarin*
	Red-handed tamarin	– *midas*
Subgenus	Moustached tamarins	*Saguinus (Tamarinus)*
Species	Black and red tamarin	– *nigricollis*
	Brown-headed tamarin	– *fuscicollis*
	White-lipped tamarin	– *weddeli*
	Golden-mantled tamarin	– *tripartitus*
	Red-mantled tamarin	– *illigeri*
	White-mantled tamarin	– *melanoleucus*
	Moustached tamarin	– *mystax*
	Lönnberg's tamarin	– *pluto*
	Rio Napo tamarin	– *graellsi*
	Red-capped tamarin	– *pileatus*
	Red-bellied tamarin	– *labiatus*
	Imperial tamarin	– *imperator*
Subgenus	Pied tamarins	*Saguinus (Marikina)*
Species	Pied tamarin	– *bicolor*
Genus	Crested or Bare-faced tamarins	*Oedipomidas*
Species	Cotton-head tamarin	– *oedipus*
	Panama wigged tamarin	– *geoffroyi*
	White-footed tamarin	– *leucopus*

Infraorder	Old World or Thin-nosed simian primates	Catarrhina
Superfamily	Old World monkeys	Cercopithecoidea
Family	Cercopithecidae	Cercopithecidae

Genus	Macaques	*Macaca (Macaca)*
Species	Magot or Barbary ape	– *sylvana*
Subgenus	Lyssodes	*Macaca (Lyssodes)*
Species	Stump-tailed macaque	– *arctoides*
	Japanese macaque	– *fuscata*
Subgenus	Rhesus monkeys	*Macaca (Rhesus)*
Species	Rhesus monkey	– *mulatta*
	Assam rhesus monkey	– *assamensis*
	Formosa macaque or Formosa rhesus	– *cyclopis*
Subgenus	Silenus	*Macaca (Silenus)*
Species	Wanderoo or Lion-tailed macaque	– *silenus*
	Pig-tailed macaque	– *nemestrina*
Subgenus	Bonnet monkeys	*Macaca (Zati)*
Species	Indian bonnet monkey	– *radiata*
	Ceylon toque monkey	– *sinica*
Subgenus	Crab-eating macaques	*Macaca (Cynomolgus)*
Species	Crab-eating macaque	– *irus*
Subgenus	Celebes macaques	*Macaca (Gymnopyga)*
Species	Moor macaque	– *maura*
Genus	Crested macaques	*Cynopithecus*
Species	Crested macaque	– *niger*
Genus	Baboons	*Papio*
Species	Chacma baboon	– *ursinus*
	Yellow baboon	– *cynocephalus*
	Anubis or Olive baboon	– *anubis*
	Guinea baboon	– *papio*
	Hamadryas or Sacred baboon	– *hamadryas*
Genus	Mandrills	*Mandrillus*
Species	Mandrill	– *sphinx*
	Drill	– *leucophaeus*
Genus	Gelada baboons	*Theropithecus*
Species	Gelada baboon	– *gelada*
Genus	Mangabeys	*Cercocebus*
Species	Sooty mangabey	– *torquatus*
	Gray-cheeked mangabey	– *albigena*
	Black mangabey	– *aterrimus*
	Agile mangabey	– *galeritus*
Genus	Guenons	*Cercopithecus*
Species	Vervet or Grass monkey	– *aethiops*
	Mona monkey	– *mona*
	Crowned guenon	– *pogonias*
	Diademed guenon	– *mitis*
	L'Hoest's monkey	– *l'hoesti*
	Diana monkey	– *diana*

Species	De Brazza's monkey	– *neglectus*	
	Lesser white-nosed guenon	– *petaurista*	
	Greater white-nosed guenon	– *nictitans*	
	Red-bellied guenon	– *erythrogaster*	
	Moustached guenon	– *cephus*	
	Hamlyn's or Owl-faced monkey	– *hamlyni*	
Subgenus	Guenons	*Cercopithecus (Miopithecus)*	
Species	Talapoin monkey	– *talapoin*	
Genus	Guenons	*Cercopithecus*	
Species	Swamp monkey	– *nigroviridis*	
Genus	Patas or Red monkeys	*Erythrocebus*	
Species	Patas or Red monkey	– *patas*	

Family	Leaf monkeys	Colobidae

Genus	Langurs	*Presbytis*
Subgenus	Langurs	*Presbytis (Semnopithecus)*
Species	Hanuman or Entellus monkey	– *entellus*
Subgenus	Purple-faced langurs	*Presbytis (Kasi)*
Species	John's langur	– *johni*
	Purple-faced langur	– *senex*
Subgenus	Capped langurs	*Presbytis (Trachypithecus)*
Species	Silvered leaf monkey	– *cristatus*
	Capped langur	– *pileatus*
	Phayre's leaf monkey	– *phayrei*
	Dusky leaf monkey	– *obscurus*
	Francois' leaf monkey	– *francoisi*
	White-headed leaf monkey	– *leucocephalus*
	Mentawai island langur	– *potenzani*
Subgenus	Island langurs	*Presbytis (Presbytis)*
Species	Maroon leaf monkey	– *rubicundus*
	Banded leaf monkey	– *melalophus*
	White-fronted leaf monkey	– *frontatus*
	Sunda Island leaf monkey	– *aygula*
Genus	Douc langurs	*Pygathrix*
Species	Douc langur	– *nemaeus*
Genus	Snub-nosed monkeys	*Rhinopithecus*
Species	Golden monkey	– *roxellanae*
	Brown snub-nosed monkey	– *bieti*
	White-mantled snub-nosed monkey	– *brelichi*
	Tonkin snub-nosed monkey	– *avunculus*

Genus	Pageh Pig-tailed monkeys	*Simias*
Species	Pageh Pig-tailed monkey	– *concolor*
Genus	Proboscis monkeys	*Nasalis*
Species	Proboscis monkey	– *larvatus*
Genus	Colobus monkeys	*Colobus*
Subgenus	Colobus monkeys	*Colobus (Procolobus)*
Species	Olive colobus monkey	– *verus*
Subgenus	Colobus monkeys	*Colobus (Piliocolobus)*
Species	Red colobus monkey	– *badius*
Subgenus	Black and white colobus monkeys or Guerezas	*Colobus (Colobus)*
Species	Northern black and white guereza	– *abyssinicus*
	Southern black and white guereza	– *polykomos*

Superfamily	Apes and men	Hominoidea
Family	True Gibbons	Hylobatidae
Genus	Siamangs	*Symphalangus*
Species	Siamang	– *syndactylus*
	Dwarf siamang	– *klossi*
Genus	True Gibbons	*Hylobates*
Species	Black gibbon	– *concolor*
	Hoolock	– *hoolock*
	White-handed gibbon or Lar	– *lar*
	Agile gibbon	– *agilis*
	Silvery gibbon	– *moloch*

Family	Great apes	Pongidae
Genus	Orang utans	*Pongo*
Species	Orang Utan	– *pygmaeus*
Genus	Gorillas	*Gorilla*
Species	Gorilla	– *gorilla*
Genus	Chimpanzees	*Pan*
Species	Chimpanzee	– *troglodytes*
	Bonobo or Dwarf chimpanzee	– *paniscus*

Family	Men	Hominidae
Genus	Men	*Homo*
Species	Man	– *sapiens*

Subgenera (scientific names in brackets) are included. Subspecies are not listed here.
Based on the classification in *Grzimeks Tierleben* (Kindler Verlag, Zurich, 1967)

Bibliography

ALBRECHT, H., and S.C.DUNNET: *Chimpanzees in Western Africa*. Munich, 1971.

ALTMANN, J.: *Baboon Mother and Infants*. Cambridge, Mass., London, 1980.

ALTMANN, S.A.: *Social Communication among Primates*. Chicago, London, 1967.

ALTMANN, S.A., and J. ALTMANN: "Baboon Ecology," in: *Bibliotheca Primatologica* 7 (1970).

ALTMANN-SCHÖNBERNER, D.: "Beobachtungen über Aufzucht und Entwicklung des Verhaltens beim Grossen Löwenäffchen, Leontocebus rosalia," in: *Der Zoolog. Garten*, New Series, 31 (1965).

ANGST, W: *Aggression bei Affen und Menschen*. Berlin, Heidelberg, New York, 1980.

ANGST, W.: "Das Ausdrucksverhalten des Javaneraffen Macaca fascicularis Raffles," in: *Fortschritte der Verhaltensforschung* 15 (1974).

ANGUS, SHANNON: "Water-Contact Behavior of Chimpanzees," in: *Folia Primatologica* 14 (1971).

ANKEL, F.: *Einführung in die Primatenkunde*. Stuttgart, 1970.

ANONYMUS: "En el Parque Zoológico de Barcelona nace otro Gorila," in: *Zoo Revista del Parque Zoológico de Barcelona* 32 (1978).

ASANOW, S., and W.LIPPERT: "Bemerkungen zum Verhalten einer Mantelpaviangruppe (Papio hamadryas) gegenüber einer Mutter mit Zwillingen," in: *Der Zoolog. Garten*, New Series, 46 (1976).

ASHTON, E.H., and R.L.HOLMES: "Perspectives in primate biology," in: *Zeitschr. f. Tierpsych.* 58 (1982).

ATTENBOROUGH, D.: *Tierversuche auf Madagaskar*. Zurich, 1962.

AUTRUM, H.: *Menschliches Verhalten als biologisches Problem*. Munich, 1976.

AVELING, R., and C. AVELING: "Ein Hoffnungschimmer für die sanften Riesen," in: *Das Tier* 21 (1981), No.9.

BADRIAN, A., and N.BADRIAN: "Dem Geheimnis der Bonobos auf der Spur," in: *Grzimeks Tier, Sielmanns Tierwelt* 22 (1982), No.5.

BALDWIN, J.D.: "The Social Organization of a Semifree Ranging Troop of Squirrel Monkeys (Saimiri sciureus)," in: *Folia Primatologica* 14 (1971).

BARASH, D.P.: *Soziobiologie und Verhalten*. Berlin, Hamburg, 1980.

BARNETT, S.A.: *"Instinct" and "Intelligence," the Science of Behaviour in Animals and Man*. London, 1967.

BASILEWSKI, G.: "Haltung und Zucht der Madagaskar Makis in Gefangenschaft," in: *Freunde des Kölner Zoo* 8 (1965), No.1.

BAUMGÄRTEL, W.: *König im Gorillaland*. Stuttgart, 1960.

BAUMGÄRTEL, W.: *Unter Gorillas*. Berlin, 1977.

BERGER, G.: "Backenwülster—Zum Aussterben verurteilt?," in: *Wissenschaft und Fortschritt* 13 (1963).

BERGER, G.: "Künstliche Aufzucht von Tupaias (Tupaia tana)," in: *Freunde des Kölner Zoo* 9 (66/67), No.4.

BERGER, G.: "Orang-Utans," in: *Priroda* 1, (1966).

BERNSTEIN, I.S.: "Activity Patterns in a Gelada Monkey Group," in: *Folia Primatologica* 23 (1975).

BERNSTEIN, I.S.: "A Field Study of the Pigtail Monkey (Macaca nemestrina)," in: *Primates* 8 (1977).

BERNSTEIN, I.S.: "The Lutong of Kuala Selangar," in: *Behavior* 26 (1968).

BERTRAND, M.: "The behavioural Repertoire of the Stumptail Macaque," in: *Bibliotheca Primalologica* 11 (1969).

BLAFER-HARDY, S.: *The Langurs of Abu*. Cambridge, 1977.

BOCK, D.: "Die letzten Berggorillas," in: *Der Zoolog. Garten*, New Series, 1 (1928).

BOETTICHER, H. VON: *Die Halbaffen und Koboldmakis. Neue Brehm-Bücherei*. Wittenberg Lutherstadt, 1958.

BÖTTCHER, A.R.: *Die Affensache*. Berlin, no date.

BORNER, M.: "Rubondo, ein Nationalpark mausert sich," in: *Das Tier* 20 (1980), No. 10.

BORNER, M., and B.STONEHOUSE: *Orang-Utan, Orphans of the Forest*. London, 1979.

BORNER-LOEWENSBERG, M.: "Ein Orang wird vom Baum geholt," in: *Das Tier* 17 (1977), No.2.

BOURNE, G.H.: *Das Volk der Affen*. Munich, 1971.

BOURNE, G.H.: "Die Affen von Gibraltar," in: *Das Tier* 20 (1980), No. 12.

BOURNE, G.H., and M.COHEN: *Die sanften Riesen*. Munich, 1977.

BRANDES, G.: *Buschi. Vom Orang-Säugling zum Backenwülster*. Leipzig, 1939.

BRANDES, G.: "Die Menschenaffen," in: *Mitteil. aus dem Zoolog. Garten Dresden*, 1912.

Breeding Primates. International Symposium on Breeding Non-human Primates for Laboratory Use. Editor: W.I.B.Beveridge. Berne, 1971; Basle, Munich, Paris, London, New York, Sydney, 1972.

Brehms Neue Tierenzyklopädie. Säugetiere. Vol.2, Freiberg, Basle, Vienna, 1978.

Brehms Tierleben. Editor: Otto zur Strassen. *Säugetiere*, 4. Leipzig, 1922.

BRENTJES, B.: "Riesen des Altertums—Gorgonen und Gorillas," in: *Der Zoodirektor erzählt*, 16/1964.

BREUER, G.: "Mann und Frau bei Tier und Mensch," in: *Naturwiss. Rundschau* 32 (1979).

BREWER, S.: *Die Affenschule. Neue Wege der Wildtierforschung*. Vienna, Hamburg, 1978.

BROWN, K., and D.S.MACK: "Food Sharing among Captive Leontopithecus rosalia," in: *Folia Primatologica* 29 (1978).

BUCHHOLTZ, C.: *Das Lernen bei Tieren*. Stuttgart, 1973.

BUDNITZ, N., and K.DAINIS: "Lemur catta: Ecology and Behavior," in: *Lemur Biology* (1975).

BÜRGER, M., U.SEDLAG and R.ZIEGER: *Zooführer*. Leipzig, Jena, Berlin, 1980.

CAMPBELL, B.G.: *Entwicklung zum Menschen*. Stuttgart, 1972.

CARRINGTON, R.: *Säugetiere. Time-Life International* (Nederland) N.V., 1965.

CHALMERS, N.R.: "The visual and vocal communication of free-living mangabeys in Uganda," in: *Folia Primatologica* 9 (1968).

CHANCE, M.R.A., and C.J.JOLLY: *Social Groups of Monkeys, Apes and Men*. London, 1970.

CHIARELLI, A.B.: *Taxonomy Atlas of Living Primates*. London, New York, 1972.

CHIVERS, D.J.: "Communication within and between family groups of Siamang (Symphalangus syndactylus)," in: *Behavior* 57 (1976).

CHIVERS, D.J.: *The Siamang in Malaya. A Field Study of a Primate in Tropical Rain Forest. Contributions to Primatology*, 4. Basle, Munich, Paris, London, New York, Sydney, 1974.

CHIVERS, D.J., J.J.RAEMAEKERS, and F.P.G.ALDRICH-BLAKE: "Long-term Observations of Siamang Behavior," in: *Folia Primatologica* 23 (1975).

CLOCHON, R.L., and A.B.CHIARELLI (Editors): *Evolutionary Biology of the New World Monkeys and Continental Drift*. New York, London, 1980.

CLUTTON-BROCK, T.H. (Editor): *Primate Ecology*. London, New York, San Francisco, 1977.

COE, C.L., and L.A.ROSENBLUM: "Annual Reproductive Strategy of the Squirrel Monkey (Saimiri sciureus)," in: *Folia Primatologica* 29 (1978).

COIMBRA-FILHO, A.F., and A.R.MITTERMEIER (Editors): *Ecology and Behaviour of Neotropical Primates*. Vol. I, Rio de Janeiro, 1981.

COUSINS, D.: "The Breeding of Gorillas, Gorilla gorilla, in Zoological Collections," in: *Der Zoolog. Garten*, New Series, 46 (1976).

DATHE, R.: "Familienähnlichkeiten bei den Sumatra-Orang-Utans, Pongo pygmaeus abeli, im Tierpark Berlin," in: *Milu* 5 (1980), No. 1/2.

DATHE, R.H., H.DATHE and R.NAGEL: "Beobachtung zur Mutter-Kind-Beziehung beim Orang-Utan (Pongo pygmaeus)," in: *Der Zoolog. Garten*, New Series, 46 (1976).

DEMBROWSKI, J.: *Die Psychologie der Affen*. Berlin, 1956.

DE VORE, I., and S.EIMERL: *Die Primaten. Time-Life International* (Nederland) N.V. 1966.

DE WAAL, F.B.M.: "Exploitative and familiarity-dependent support strategies in a colony of semi-free living Chimpanzees," in: *Behavior* 66 (1978).

DIETRICH, G.: "Affen helfen bei der Ernte," in: *Das Tier* 20 (1980), No.4.

DITTRICH, L.: "Beobachtungen an freigekommenen Guerezaaffen (Colobus polykomos caudatus Thomas 1885)," in: *Der Zoolog. Garten*, New Series, 32 (1966).

DITTRICH, L.: "Tupaias im Tropenhaus," in: *Der Zoofreund* 24 (1977).

Die Welt der wilden Tiere. Affen. Munich, 1981.

DIXSON, A.F.: *The Natural History of the Gorilla*. New York, 1981.

DOBZHANSKY, T.: *Die Entwicklung zum Menschen*. Hamburg, Berlin, 1958.

DÖHL, J.: "Über die Fähigkeit einer Schimpansin, Umwege mit selbständigen Zwischenzielen zu überblicken," in: *Zeitschr. f.Tierpsych.* 25 (1968).

DÖHL, J., and D.PODOLCZAK: "Versuche zur Manipulierfreudigkeit von zwei jungen Orang-Utans (Pongo pygmaeus) im Frankfurter Zoo," in: *Der Zoolog. Garten*, New Series, 43 (1973).

DORST, J., and P.DANDELOT: *Säugetiere Afrikas*. Hamburg, Berlin, 1970.

DOYLE, G.A., and S.K.BEARDER: *The Galagos of South Africa. Primate Conservation*. New York, San Francisco, London, 1977.

DUNBAR, R.J.M., and M.F.NATHAN: "Social Organization of the Guinea Baboon, Papio papio," in: *Folia Primatologica* 17 (1972).

EDEY, M.A.: *Vom Menschenaffen zum Menschen. Time-Life International* (Nederland) N.V., 1975.

EIBL-EIBESFELDT, I.: *Krieg und Frieden*. Munich, Zurich, 1975.

EIMERL, S., and I. DE VORE: *Die Primaten. Time-Life International* (Nederland) N.V., 1966.

EPPLE, G.: "Vergleichende Untersúchungen über Sexual- und Sozialverhalten der Krallenaffen (Hapalidae)," in: *Folia Primatologica* 8 (1967).

EWER, R.F.: *Ethologie der Säugetiere*. Hamburg, Berlin, 1976.

FÄDRICH, H., and J.FÄDRICH: *Zooführer Säugetiere*. Stuttgart, 1973.

Fauna. Vols. II, III, VII, IX, Munich, 1971.

FIEDLER, W.: "Übersicht über das System der Primaten," in: *Primatologia* I (1956).

FISCHER, F.: *Das Jahr mit den Gibbons*. Wittenberg Lutherstadt, 1965.

FISCHER, W.: "Einige Ergänzungen zur Haltung und Entwicklung des Schopfgibbons Hylobates (Nomascus) concolor (Harlan)," in: *Milu* 5 (1980), No. 1/2.

FOSSEY, D.: "Making Friends with Mountain Gorillas," in: *National Geograph. Mag.* 137 (1970), No. 1.

FOSSEY, D.: "More Years with Mountain Gorillas," in: *National Geograph. Mag.* 140 (1971).

FRANTZ, J.: "Beobachtungen bei einer Löwenäffchen-Aufzucht," in: *Der Zoolog. Garten*, New Series, 28 (1963).

FREEMAN, D.: *The Great Apes*. London, New York, Sydney, Toronto, 1979.

GALDIKAS-BRINDAMOUR, B.: "Orang utans, Indonesia's 'People of the Forest,'" in: *National Geograph. Mag.* 148 (1975), No.4.

GARDNER, R.A., and B.T.GARDNER: "Teaching sign language to a chimpanzee," in: *Science* 165 (1969).

GAUTIER-HION, A.: "L'organisation sociale d'une bande de Talapoins (Miopithecus talapoin) dans le Nord-Est de Gabun," in: *Folia Primatologica* 12 (1970).

GAUTIER-HION, A., and J.P.GAUTIER: "Croissance maturité sexuelle et sociale reproduction chez les cercopithécines forestier africains," in: *Folia Primatologica* 26 (1976).

GAUTIER-HION, A., and J.P.GAUTIER: "Le Singe de Brazza: Une stratégie originale," in: *Zeitschr. f. Tierpsych.* 46 (1978).

GEDDES, H.: *Gorilla*. Wiesbaden, 1956.

GEHRING, C.B.: "Grussverhalten und Erkennen vertrauter Personen sowie weitere Verhaltensweisen einer Kapuziner-Gruppe (Cebus apella) im Zoo," in: *Der Zoolog. Garten*, New Series, 46 (1976).

GENSCH, W.: "Einige Bemerkungen zum Wachstum des weiblichen Gorillas 'Dima' im Zoologischen Garten Dresden," in: *Der Zoolog. Garten*, New Series, 32 (1966).

GEORG, M: "Baden ja, schwimmen nein," in: *Das Tier* 20 (1980), No.5.

GERSTER, G.: "Sumatra: Kleine Nebelparder töten Orang-Utans!," in: *Das Tier* 15 (1975), No.4.

GEWALT, W.: *Bahala. Ein Gorilla lebt in unserer Küche*. Stuttgart, 1964.

GORGAS, M.: "Berggorillas—eine Aufgabe für Zoo und Naturschutz," in: *Freunde des Kölner Zoo* 12 (1969), No.3.

GORGAS, M.: "Geschlechtsbestimmung bei Gorillas mit der Haarfollikel-Methode ('Olympiatest')," in: *Zeitschr. d. Kölner Zoo* 16 (1973), No. 3.

GORGAS, M.: "Zur Problematik der Aufzucht von Orang-Utans im Zoo," in: *Zeitschr. d. Kölner Zoo* 15 (1972), No. 3.

GOVAN, J. A.: *The Non-Human Primates and Human Evolution.* Detroit, 1957.

GRAHAM, CHARLES. E.: *Reproductive Biology of the Great Apes: Comparative and Biomedical Perspectives.* New York, 1981

GRONEFELD, G.: "Das Haus der Urwaldgeister," in: *Das Tier* 17 (1977), No. 7.

GRONEFELD, G.: "Roboter ersetzen Menscheneltern. Neue Wege bei der Aufzucht von Lisztaffen," in: *Das Tier* 19 (1979), No. 12.

GRONEFELD, G.: *Verstehen wir die Tiere?* Brunswick, 1963.

GROVES, C. P.: *Gorillas. The World of Animals.* London, 1970.

GRZIMEK, B.: *Flug ins Schimpansenland.* Stuttgart, 1952.

GRZIMEK, B.: "Schreiend raste der Gorilla auf mich zu," in: *Das Tier* 14 (1974), No. 7.

Grzimeks Tierleben. Vols. X and XI: *Herrentiere (Primaten).* Zurich, 1968.

GSCHWEND, J.: "Ich blickte wilden Berggorillas in die Augen," in: *Das Tier* 12 (1972), No. 6.

GUCWINSKI, A.: "Acclimatization of Young Lowland Gorillas (Gorilla g. gorilla) in Wrocław Zoological Garden during the Years 1970–1978," in: *Der Zoolog. Garten,* New Series, 51 (1981).

HALL, K. L. R.: *Variations in the Ecology of the Chacma Baboon, Papio ursinus. The Primates. Symposia of the Zoological Society of London,* 10 (1963).

HALTENORTH, T., and H. DILLER: *Säugetiere Afrikas und Madagaskars.* Munich, Berne, Vienna, 1977.

Handgebrauch und Verständigung bei Affen und Frühmenschen. Editor: B. Rensch. Berne, Stuttgart, 1968.

HANSEN, H.: "Sozial- und Territorialverhalten des Senegalgalagos (Galago senegalensis)," in: *Zeitschr. d. Kölner Zoo* 21 (1978/79), No. 3.

HARCOURT, A. H., and K. J. STEWART: "Chimpanzee, Gorilla and Man," in: *New Scientist,* Vol. 76, 1977.

HARRINGTON, J.: "Field Observations of Social Behavior of Lemur f. fulvus," in: *Lemur Biology* (1975).

HARRINGTON, J. E.: "Diurnal Behavior of Lemur mongoz at Ampijoroa, Madagaskar," in: *Folia Primatologica* 29 (1978).

HARRISON, B.: *Education to Wild Living of Young Orang-Utans at Bako National Park, Sarawak.* No place, no date.

HARRISSON, B.: *Kinder des Urwalds.* Wiesbaden, 1964.

HARRISSON, B.: *Orang-Utan.* London, 1962.

HASSELBACHER, W.: "Ein Orang Utan, namens Karl," in: *Das Tier* 22 (1982), No. 1.

HAUSFATER, G.: "Predatory behavior of Yellow Baboons," in: *Behavior* 56 (1976).

HAUSWIRTH, H.: *Hanuman. Eine Erzählung von den Heiligen Affen Indiens.* Erlenbach-Zurich, Leipzig, no date.

HAYES, C.: *The Ape in our House.* New York, 1951.

HEBERER, G., W. HENKE, and H. ROTHE: *Der Ursprung des Menschen.* Stuttgart, 1975.

HECK, H.: *Die Hellabrunner Schimpansenzucht. Das Tier und wir.* Munich, 1939.

HEDIGER, H.: "Ausverkauf der letzten Affen," in: *Das Tier* 16 (1976), No. 7.

HEDIGER, H.: "Sprechen Sie yerkisch? Die neueste Umgangssprache mit Schimpansen," in: *Das Tier* 19 (1979), No. 1.

HEDIGER, H.: "Vierbeinige Schelme. Auch Tiere haben Humor," in: *Das Tier* 19 (1979), No. 5.

HEMMER, H.: "Beobachtungen zur Sozialbiologie madagassischer Lemuren," in: *Zeitschr. d. Kölner Zoo* 22 (1979), No. 2.

HICK, U.: "Aus der Kölner Lemuren-Sammlung," in: *Freunde des Kölner Zoo* 10 (1967), No. 2.

HICK, U.: "Aus der Kölner Saki-Sammlung," in: *Freunde des Kölner Zoo* 9 (1966), No. 3.

HICK, U.: "Das erste Jahr im neuen Lemurenhaus," in: *Zeitschr. d. Kölner Zoo* 17 (1974), No. 4.

HICK, U.: "Einige Beobachtungen zur Fortpflanzung und Jugendentwicklung von Katta (Lemur catta), Mongozmaki (Lemur mongoz mongoz), Kronenmaki (Lemur mongoz coronatus) und Weisskopfmaki (Lemur fulvus albifrons)," in: *Freunde des Kölner Zoo* 12 (1969), No. 3.

HICK, U.: "Erstmalige gelungene Zucht eines Bartsakis im Kölner Zoo," in: *Freunde des Kölner Zoo* 11 (1968), No. 2.

HICK, U.: "Hand-rearing a Ring-tailed Lemur, Lemur catta, and a Crowned Lemur, Lemur mongoz coronatus, at Cologne Zoo," in: *The Internat. Zoo Yearbook* 16, London, 1976.

HICK, U.: "Nasenaffen (Nasalis larvatus larvatus) im Kölner Zoo," in: *Freunde des Kölner Zoo* 9 (1966/67), No. 4.

HICK, U.: "Wir sind umgezogen," in: *Zeitschr. d. Kölner Zoo* 16 (1973), No. 4.

HICK, U.: "Zucht und Haltung von Kleideraffen (Pygathrix nemaeus Linnaeus 1771) im Zoologischen Garten Köln," in: *Freunde des Kölner Zoo* 13 (1970), No. 3.

HIDEYUKI, O.: "Verkrüppelte Affen warnen die Menschheit" (Japanese), in: *Das Tier* 19 (1979) No. 2.

HILL, W. C. O.: *Evolutionary Biology of the Primates.* London, New York, 1972.

HILL, W. C. O.: *Primates—Comparative Anatomy and Taxonomy.* Vols. 1–6, Edinburgh, 1953–1955.

HILLARY, E., and D. DOIG: *Schneemenschen und Gipfelstürmer.* Wiesbaden, 1963.

HINDE, R. A.: *Animal Behavior: A Synthesis of Ethology and Comparative Psychology.* New York, 1966.

HOFER, H.: "Über den Gesang des Orang-Utan (Pongo pygmaeus)," in: *Der Zoolog. Garten,* New Series, 41 (1971/72).

HOFER, H., and G. ALTNER: *Die Sonderstellung des Menschen.* Stuttgart, 1972.

HUXLEY, T. H.: *Zeugnisse für die Stellung des Menschen in der Natur.* Stuttgart, 1970.

ILLIES, J.: *Die Affen und wir. Ein Vergleich zwischen Verwandten.* Reinbek near Hamburg.

ITANI, J.: "On the Acquisition and Propagation of a new Food Habit in the Troop of Japanese Monkeys at Takasakiyama," in: *Primates* 1 (1958).

ITANI, J.: "The social construction of national troops of Japanese monkeys in Takasakiyama," in: *Primates* 4 (1963).

ITANI, J.: "Vocal communication in the wild Japanese monkey," in: *Primates* 4 (1963).

ITANI, J., and A. SUZUKI: "The social unit of chimpanzees," in: *Primates* 8 (1967).

IZAWA, K., and T. NISHIDA: "Monkeys living in the northern limit of their distribution," in: *Primates* 4 (1963).

JACKSON, P.: "Die Schimpansenschule," in: *Das Tier* 20 (1980), No. 9.

JANTSCHKE, F.: "Affenliebe kann auch schädlich sein. Wie Orang-Utans ihre Babys erziehen," in: *Das Tier* 20 (1980), No. 8.

JANTSCHKE, F.: "Die Schneeaffen vom Höllental," in: *Das Tier* 17 (1977), No. 4.

JANTSCHKE, F.: *Orang-Utans in zoologischen Gärten*. Munich, 1972.

JANTSCHKE, F.: "Schade um den schönen Gorilla," in: *Das Tier* 15 (1975), No. 1.

JANTSCHKE, F.: "Schimpansen bleiben voller Rätsel," in: *Das Tier* 19 (1979), No. 2.

JANTSCHKE, F.: "Sind Paviane gar nicht so gefährlich?," in: *Das Tier* 20 (1980), No. 6.

JANTSCHKE, F.: "Wie alt werden Affen?," in: *Das Tier* 18 (1978), No. 4.

JAY, P.: "The common langur of North India," in: *Primate Behavior* (1965).

JOHNSON, M.: *Congorilla*. Leipzig, 1940.

JOHST: *Biologische Verhaltensforschung am Menschen*. Berlin, 1976.

JOLLY, A.: *A World like our Own. Man and Nature in Madagascar*. New Haven, London, 1980.

JOLLY, A.: *Lemur Behavior*. Chicago, London, 1966.

JONES, C., and S. J. PI: "Comparative Ecology of Gorilla gorilla (Savage and Wyman) and Pan troglodytes (Blumenbach) in Rio Mundi, West Africa," in: *Bibliotheca Primatologica* 13 (1971).

JONES, C., and S. J. PI: "Comparative Ecology of Cercocebus albigena (Gray) and Cercocebus torquatus (Ker) in Rio Mundi, West Africa," in: *Folia Primatologica* 9 (1968).

JONES, M. L.: *Studbook of the Orang Utan Pongo pygmaeus*. San Diego, 1980.

JUNTKE, C.: "Über die künstliche Aufzucht und Entwicklung von 5 Schimpansengeschwistern. Vorbericht von M. Bürger," in: *Der Zoolog. Garten*, New Series, 50 (1980).

JUPPENPLATZ, P.: "Die Dschungelakrobaten (Gibbons)," in: *Der Stern*, Hamburg, 1980.

KAINZ, F.: *Die "Sprache" der Tiere*. Stuttgart, 1961.

KAWABE, M.: "One observed case of hunting behavior among wild chimpanzees living in the savanna woodland of Western Tanzania," in: *Primates* 7 (1966).

KAWAI, M.: "Ecological and sociological Studies of Gelada Baboons," in: *Contributions to Primatology* 16 (1979).

KAWAI, M.: "Newly acquired precultural behavior of the natural troop of Japanese monkeys on Koshima Islands," in: *Primates* 6 (1965).

KAWATA, B.: "Notes on Comparative Behavior in Three Primate Species in Captivity," in: *Der Zoolog. Garten*, New Series, 50 (1980).

KETTER, M. D. and L. P. PICHETTE: "Reproductive Behavior of Captive Subadult Lowland Gorillas (Gorilla g. gorilla)," in: *Der Zoolog. Garten*, New Series, 49 (1979).

KIRCHSHOFER, R.: "Die erste Geburt eines Zwergschimpansen in einem Zoo," in: *Umschau* 62 (1962).

KIRCHSHOFER, R.: "Gorillazucht in zoologischen Gärten und Forschungsstationen," in: *Der Zoolog. Garten*, New Series, 38 (1970).

KIRCHSHOFER, R.: "Stammesgeschichte der Primaten," in: *Mitteil. aus dem Frankfurter Zoo* 33 (1978).

KLEEMANN, G.: *Die peinlichen Verwandten*. Stuttgart, 1966.

KLÖS, U.: "Kann der Orang-Utan (Pongo pygmaeus L.) in der Freiheit überleben?," in: *Bongo* (1979), No. 3.

KLOSE H., and I. KLOSE: "Gorilla Twins Gorilla g. gorilla born at Frankfurt Zoo," in: *International Yearbook* 8, London, 1968.

KÖHLER, W.: *Intelligenzprüfungen an Menschenaffen*. Berlin, Göttingen, Heidelberg, 1963.

KOLAR, K.: "Einige Mitteilungen über Fortpflanzung und Jugendentwicklung von Galago senegalensis in Gefangenschaft," in: *Der Zoolog. Garten*, New Series 31 (1965).

KORTLANDT, A.: *Experimentation with Chimpanzees in the Wild*. Stuttgart, 1967.

KORTLANDT, A.: *New Perspectives on Ape and Human Evolution*. Amsterdam, 1972.

KORTLANDT, A., and M. KOOIJ: *Protohominid Behavior in Primates (Preliminary Communication). The Primates* (Symp. Zoolog. Soc., London, No. 10), 1963.

KRIEG, H: *Zwischen Anden und Atlantik*. Munich, 1948.

KRISHNAN, M.: "An Ecological Survey of the larger Mammals of Peninsula India (Monkeys)," in: *Journal of the Bombay Natural History Society* 68 (1971).

KRÜGER, G. R. F.: "Coco und Pucker, die Berggorillas (Gorilla g. beringei) des Kölner Zoo—ein Epilog," in: *Zeitschr. d. Kölner Zoo* 22 (1979), No. 3.

KUHN, H.-J.: "Die Geschichte der Säugetiere Madagaskars," in: *Zeitschr. d. Kölner Zoo* 15 (1972), No. 1.

KUMMER, H.: *Social Organisation of Hamadryas Baboons. A field study*. Basle, New York, 1968.

LAIDLER, K.: *The Talking Ape*. London, 1980.

LAMBRECHT, J.: "Duettgesang beim Siamang Symphalangus syndactylus," in: *Zeitschr. f. Tierpsych.* 27 (1970).

LANG, E. M.: "Ein Colobus-Bastard," in: *Der Zoolog. Garten*, New Series, 43 (1973).

LANG, E. M., R. SCHENKEL, and E. SIEGRIST: *Gorilla—Mutter und Kind*. Basle, Hamburg, Vienna, 1965.

LASLEY, B. L., J. F. KENNEDY, P. T. ROBINSON, M. H. BOGART, and K. BENIRSCHKE: "A Study of Reproductive Failure in a Pygmy Chimpanzee," in: *Der Zoolog. Garten*, New Series, 47 (1977).

LAWICK-GOODALL, J. VAN: "Chimpanzees," in: *National Geograph. Mag.* 155 (1979).

LAWICK-GOODALL, J. VAN: "Infant Killing and Cannibalism in Free-living Chimpanzees," in: *Folia Primatologica* 28 (1977).

LAWICK-GOODALL, J. VAN: "My Life among Wild Chimpanzees," in: *National Geograph. Mag.*, 130 (1963).

LAWICK-GOODALL, J. VAN: "Schimpansen erbeuten Menschenkinder," in: *Das Tier* 12 (1972), No. 7.

LAWICK-GOODALL, J. VAN: *Wilde Schimpansen*. Reinbek near Hamburg, 1971.

LEMMON, R. S.: *All about Monkeys*. New York, Toronto, 1958.

LINDEN, E.: *Die Kolonie der sprechenden Schimpansen*. Vienna, Munich, 1980.

LIPPERT, W.: "Beobachtungen zum Schwangerschafts- und Geburtsverhalten beim Orang-Utan (Pongo pygmaeus) im Tierpark Berlin," in: *Folia Primatologica* 21 (1974).

LIPPERT, W.: "Erfahrungen bei der Aufzucht von Orang-Utans (Pongo pygmaeus) im Tierpark Berlin," in: *Der Zoolog. Garten*, New Series, 47 (1977).

LORENZ, K.: *Antriebe tierischen und menschlichen Verhaltens*. Munich, 1971.

Lorenz, K.: *Das sogenannte Böse. Zur Naturgeschichte der Aggression.* Vienna, 1963.

Lorenz, K.: "Die angeborenen Formen möglicher Erfahrung," in: *Zeitschr. f. Tierpsych.* 5 (1943).

Lorenz, K.: *Vergleichende Verhaltensforschung.* Vienna, New York, 1979.

Lorenz, R.: "Zur Haltung des Springtamarins Callimico goeldii (Thomas 1904) in Deutschland," in: *Der Zoolog. Garten,* New Series, 32 (1966).

Love, J.A.: "A Note on the Birth of a Baboon (Papio anubis)," in: *Folia Primatologica* 29 (1978).

Luckett, W.P., and F.S.Szalay: *Phylogeny of the Primates (a Multidisciplinary Approach).* New York, London, 1975.

Malinow, M.R.: "Biology of the Howler Monkey (Alouatta caraya)," in: *Bibliotheca Primatologica* 7 (1968).

Marler, P., and W.J.Hamilton: *Tierisches Verhalten.* Berlin, 1972.

Marsh, C.W.: "Comparative Aspects of Social Organization in the Tana River Red Colobus, Colobus badius rufomitratus," in: *Zeitschr. f. Tierpsych.* 51 (1979).

Martin, R.D.: *Breeding endangered Species in Captivity.* London, New York, San Francisco, 1975.

McKinnon, J.: *Auf der Suche nach dem roten Affen.* Zurich, 1974.

Menzel, E.W.jr., D.Premack, and G.Woodruff: "Map Reading by Chimpanzees," in: *Folia Primatologica* 29 (1978).

Merfield, F.G., in collaboration with H.Miller: *Gorillas were my Neighbours.* London, New York, Toronto, 1956.

Milton, K.: *The Foraging Strategy of Howler Monkeys.* New York, 1980.

Mizuhara, H.: "Social Changes of Japanese Monkey Troops in Takasakiyama," in: *Primates* 4 (1964).

Montagna, W.: *Non-human Primates in Biomedical Research.* Minneapolis, 1976.

Morris, D.: *Der Menschenzoo.* Munich, 1972.

Morris, D.: *Der nackte Affe.* Munich, Zurich, 1968.

Morris, D., and R. Morris: *Men and Apes.* London, 1966.

Morris, R., and D.Morris: *Der grosse Affenspiegel (eine Kulturgeschichte der Affen).* Munich, 1970.

Moynihan, M.: *The New World Primates.* Princeton, N.J., 1976.

Nadler, R.D., and B.L.Tilford: "Agonistic Interactions of Captive Female Orang Utans with Infants," in: *Folia Primatologica* 28 (1977).

Napier, J.R., and P.H.Napier: *A Handbook of Living Primates.* London, New York, 1967.

Napier, J.R., and P.H.Napier: *Old World Monkeys: Evolution, Systematics and Behavior.* New York, 1970.

Nash, L.T.: "The development of the Mother-Infant Relationship in Wild Baboons (Papio anubis)," in: *Animal Behavior* 26 (1978).

Neugebauer, W.: "The Status and Management of the Pygmy Chimpanzee Pan paniscus in European Zoos," in: *Internat. Zoo Yearbook* 20, London, 1980.

Neville, M.K.: "The Population Structure of Red Howler Monkeys (Alouatta seniculus) in Trinidad and Venezuela," in: *Folia Primatologica* 17 (1972).

Oates, J.F.: "The Social Life of a Black-and-white Colobus Monkey, Colobus guereza," in: *Zeitschr. f. Tierpsych.* 45 (1975).

Osborn, R.M.: *Observations on the Behaviour of the Mountain Gorilla. The Primates* (Symposia of the Zoological Society of London, No. 10), 1963.

Parker, C.E.: "Bob Orang Utan, from a Sarawak Jungle to a College Education," in: *ZOONOOZ XLII* (1969), No.3.

Patterson, F.: "Conversations with a Gorilla," in: *National Geograph. Mag.* 154 (1978), No.4.

Peters, G.: "Die Lemurensammlung des Zoo Tananarive, Madagaskar," in: *Zeitschr. d. Kölner Zoo* 18 (1975), No.3.

Petter, J.-J.: *The Aye-aye. Primate Conversation.* New York, San Francisco, London, 1977.

Petter, J.-J.: "The Lemurs of Madagascar," in: I. De Vore: *Primate Behavior, Field Studies of Monkeys and Apes,* New York, 1965.

Petter, J.-J., and A.Payrieras: "Preliminary Notes on the Behavior and Ecology of Hapalemur griseus," in: *Lemur Biology* (1975).

Petter, J.-J., A.Schilling, and G.Pariente: "Observations on Behavior and Ecology of Phaner furcifer," in: *Lemur Biology* (1975).

Poglayen-Neuwall, I.: "Zur Fortpflanzungsbiologie und Jugendentwicklung von Potos flavus (Schreber 1774)," in: *Der Zoolog. Garten,* New Series, 46 (1976).

Poglayen-Neuwall, J., and J.Poglayen-Neuwall: "Brüllaffen: Können wir ihre Geheimnisse noch rechtzeitig entschleiern?," in: *Das Tier* 17 (1977), No.1.

Poirier, F.E.: "Nilgiri langur (Presbytis johnii) territorial behavior," in: *Primates* 9 (1968).

Poirier, F.E.: "The Nilgiri Langur (Presbytis johnii) Troop: Its Composition, Structure, Formation, and Change," in: *Folia Primatologica* 10 (1969).

Poirier, F.E.: "The St. Kitts Green Monkey (Cercopithecus aethiops sabaeus): Ecology, Population, Dynamics, and Selected Behavioral Traits," in: *Folia Primatologica* 17 (1972).

Pollock, J.J.: "Field Observations on Indri indri. A Preliminary Report," in: *Lemur Biology* (1975).

Pook, A.G., and G.Pook: "A Field Study of the Socio-Ecology of the Goeldi's Monkey (Callimico goeldii) in Northern Bolivia," in: *Folia Primatologica* 35 (1981).

Prater, S.H.: *Indian Animals.* Bombay, 1971.

Premack, David, and A.J.Premack: *The Mind of an Ape.* New York, 1983.

Primate Conservation. Editors: Rainier III of Monaco and G.H.Bourne. New York, 1977.

Primatologie (Handbuch der Primatenkunde). Vols. I–IV. Editors: Hofer, H., A.H. Schultz, D.Starck. Basle, Munich, Paris, London, New York, Sydney, 1956–1973.

Proceedings of the Second Internat. Congress of Primatology. Vols. 1–3.

Pustorino, F., and A.Daturi: *Die Welt der Tiere, Affen.* Freiburg, Basle, Vienna, 1978.

RAEMAEKERS, J.: "Ecology of Sympatric Gibbons," in: *Folia Primatologica* 31 (1979).

RAHAMAN, H.: "The Langurs of the Gir Sanctuary (Gujarat)—A Preliminary Survey," in: *Journal of the Bombay Natural History Society* 70 (1973), No. 2.

Recent Advances in Primatology. Vols. 1–4. Editor: D. J. CHIVERS. London, New York, San Francisco, 1978.

RENSCH, B.: *Gedächtnis, Begriffsbildung und Planhandlungen bei Tieren.* Berlin, Hamburg, 1973.

RENSCH, B.: *Homo sapiens. Vom Tier zum Halbgott.* Göttingen, 1970.

RENSCH, B., and J. DÖHL: "Spontane Aufgabenlösungen durch einen Schimpansen," in: *Zeitschr. f. Tierpsych.* 24 (1967).

RENSCH, B., and K. C. DUECKER: "Manipulierfähigkeit eines jungen Orangutans und eines jungen Gorillas. Mit Anmerkungen über das Spielverhalten," in: *Zeitschr. f. Tierpsych.* 23 (1966).

REYNOLDS, V.: *Budongo.* Wiesbaden, 1966.

RICHARD, A., and R. HEIMBUCH: "An Analysis of the Social Behavior of Three Groups of Propithecus verreauxi," in: *Lemur Biology* (1975).

RICHARD, A., and R. SUSSMAN: "Future of the Malagasy Lemurs: Conservation or Extinction?" in: *Lemur Biology* (1975).

RIOPELLE, A. J.: "Growing up with Snowflake," in: *National Geograph. Mag.* 138 (1970), No. 4.

ROSENBLUM, L. A.: *Primate Behavior.* New York, London, 1970.

ROSENBLUM, L. A., and R. W. COOPER: *The Squirrel Monkey.* New York, London, 1968.

ROTHE, H.: "Further Observations on the Delivery Behaviour of the Common Marmoset (Callithrix jacchus)," in: *Zeitschr. f. Säugetierkde.* 39 (1974).

RUBY, G.: "Ein Nasenaffe in der Kinderklinik," in: *Das Tier* 19 (1979), No. 8.

RUMBAUGH, D. M.: *Gibbon and Siamang.* Vols. 1–4. Basle, Munich, Paris, London, New York, Sydney, 1927–1976.

RUMBAUGH, D. M.: "The Birth of a Lowland Gorilla at the San Diego Zoo," in: *ZOONOOZ* 38 (1965), No. 9.

RUMBAUGH, G. M.: "The Siamang Infant, Sarah, its Growth and Development," in: *ZOONOOZ* 40 (1967), No. 3.

RYALS, D., and K. KAWATA: "Behavior of the Common Woolly Monkey, Lagothrix lagotricha, in Captivity," in: *Der Zoolog. Garten, New Series,* 49 (1979).

SÄLZLE, K.: *Tier und Mensch, Gottheit und Dämon.* Munich, Basle, Vienna, 1965.

SALZERT, W.: "In Deutschlands erstem Affenpark (Magots)," in: *Der Zoofreund* 18 (1976).

SALZERT, W.: "Selbstbewusstsein und Frechheit sind wichtiger als Körperkraft," in: *Das Tier* 16 (1976), No. 9.

SANDERSON, I. T.: *Knaurs Tierbuch in Farben. Säugetiere.* Zurich, Munich, 1956.

SANDERSON, I. T., and G. STEINBACHER: *Knaurs Affenbuch.* Zurich, 1957.

SAUER, E. G.: "Zur Biologie der Zwerg- und Riesengalagos," in: *Zeitschr. d. Kölner Zoo* 17 (1974), No. 2.

SCHALLER, G. B.: *The Mountain Gorilla.* Chicago, 1963.

SCHALLER, G. B.: *Unsere nächsten Verwandten.* Berne, Munich, Vienna, 1965.

SCHMIDBAUER, W.: "So schlau sind Schimpansen wirklich!," in: *Das Tier* 14 (1974), No. 9.

SCHMIDT, C.: "Der Affe ist los," in: *Grzimeks Tier, Sielmanns Tierwelt* 21 (1981), No. 10.

SCHMIDT, C.: "Javaneraffen auf Bali: Heiligkeit ist keine Lebensversicherung," in: *Grzimeks Tier, Sielmanns Tierwelt* 22 (1982), No. 4.

SCHNEIDER, H.-E.: "Primaten als Versuchstiere—eine verpflichtende Notwendigkeit," in: *Der Zoolog. Garten, New Series,* 49 (1979).

SCHRIER, A. M., H. F. HARLOW, and F. STOLLNITZ: *Behavior of Non-human Primates.* Vols. 1 and 2, New York, London, 1965.

SCHRIER, A. M.: "Learning—Set Formation and Transfer in Rhesus and Talapoin Monkeys," in: *Folia Primatologica* 17 (1972).

SCHULTZ, A.: "Die Primaten," in: *Die Enzyklopädie der Natur.* Lausanne, 1971.

SCHWEISHEIMER, W.: "Neue Forschungen über Zwergschimpansen," in: *Naturwissensch. Rundschau* 29 (1976), No. 8.

SCHWEISHEIMER, W.: "Zärtlichkeit ist wichtiger als Muttermilch," in: *Das Tier* 18 (1978), No. 6.

SEIFERT, S.: "Die neue Gibbonanlage im Leipziger Zoo," in: *Panthera* (1974).

SEITZ, E.: "Die Bedeutung geruchlicher Orientierung beim Plumplori (Nycticebus coucang Boddaert) (1785)," in: *Zeitschr. f. Tierpsych.* 26 (1969).

SEYFARTH, R. M., and D. L. CHANEY: "The Ontogeny of Vervet Monkey Alarm Calling Behavior," in: *Zeitschr. f. Tierpsych.* 54 (1980).

SHOEMAKER, A. H.: "Observations on Howler Monkeys, Alouatta caraya, in Captivity," in: *Der Zoolog. Garten, New Series,* 48 (1978).

SIEK, J. B.: "Patterns of Food Sharing among Mother and Infant Chimpanzees at Gombe National Park, Tanzania," in: *Folia Primatologica* 29 (1978).

SIG, H.: "Differentiation of Female Positions in Hamadryas One-Male-Units," in: *Zeitschr. f. Tierpsych.* 53 (1980).

SIMON, K. H.: "Leakey und der Ursprung des Menschengeschlechts," in: *Naturwiss. Rundschau* 29 (1976).

SMALL, M. F.: "Comparative Social Behavior of Adult Female Rhesus Macaques and Bonnet Macaques," in: *Zeitschr. f. Tierpsych.* 59 (1982).

SPRANKEL, H.: "Zucht von Tupaia glis Diard 1820 in Gefangenschaft," in: *Naturwiss. Rundschau* 47 (1960).

STARCK, D.: *Die Säugetiere Madagaskars, ihre Lebensräume und ihre Geschichte.* Wiesbaden, 1974.

STARING, E. D.: "Indiens Languren verlieren den Heiligenschein," in: *Das Tier* 21 (1981), No. 4.

STILLER, H., and M. STILLER: *Tierversuch und Tierexperiment.* Hanover, 1977.

STREBBINS, G. L.: *Evolutionsprozesse.* Jena, 1968.

STRUHSAKER, T.: "Affenkämpfe im Urwald von Kibale," in: *Das Tier* 16 (1976), No. 1.

STRUHSAKER, T.: "Infanticide and Social Organization in the Redtail Monkey (Cercopithecus ascanius schmidti) in the Kibale Forest, Uganda," in: *Zeitschr. f. Tierpsych.* 45 (1977).

STRUHSAKER, T. T.: "Correlates of Ecology and Social Organization among African Cercopithecines," in: *Folia Primatologica* 11 (1969).

STRUHSAKER, T.T.: "Social Structure among Vervet Monkeys (Cercopithecus aethiops)," in: *Behavior* 29 (1967).

SUGIYAMA, Y.: "Social Organization of Chimpanzees in the Budongo Forest, Uganda," in: *Primates* 9 (1968).

SUZUKI, A.: "An Ecological Study of Wild Japanese Monkeys in Snowy Areas focused on their Food Habits," in: *Primates* 6 (1965).

TAKAHATA, Y.: "The Socio-Sexual Behavior of Japanese Monkeys," in: *Zeitschr. f. Tierpsych.* 59 (1982).

TATTERSALL, I., and R. SUSSMAN: "Notes on Topography, Climate, and Vegetation of Madagascar," in: *Lemur Biology* (1975).

The Rhesus Monkey. E 1/2. Editor: G. H. Bourne. New York, London, 1975.

TILFORD, B. L., and R. D. NADLER: "Male Parental Behavior in a Captive Group of Lowland Gorillas (Gorilla gorilla gorilla)," in: *Folia Primatologica* 29 (1978).

TILSON, R. L.: "Family Formation Strategies of Kloss's Gibbons," in: *Folia Primatologica* 35 (1981).

TINBERGEN, N.: *Tiere und ihr Verhalten. Time-Life International* (Nederland) N. V., 1966.

TUTTLE, R.: *The Functional and Evolutionary Biology of Primates*. Chicago, New York, 1972.

UHLMANN, K.: *Der Orang-Utan—ein myologischer Aussenseiter. Verhandlungen Anatomische Gesellsch.* 67 (1973).

ULLRICH, W.: *Affen ernst genommen*. Radebeul, 1968.

ULLRICH, W.: "Das Verhalten des Schimpansen, Pan tr. troglodytes Blumenbach 1799, beim Sprung," in: *Säugetierkdl. Mitteil.* 2 (1954).

ULLRICH, W.: "Geburt und natürliche Geburtshilfe beim Orang-Utan," in: *Der Zoolog. Garten*, New Series, 39 (1970).

ULLRICH, W.: *Tiere recht verstanden*. Leipzig, Jena, Berlin, 1968.

ULLRICH, W.: "Zur Biologie und Soziologie der Colobusaffen," in: *Der Zoolog. Garten*, New Series, 25 (1961).

ULLRICH, W., and E. TYLINEK: *Endangered Species*. New York, 1972.

ULMER, F. A.: *First Orang Utan Born in Rare Mammal House. America's First Zoo*, Philadelphia, 18 (1966), No. 3.

ULMER, F. A.: *Massa—Dean of Captive Gorillas. America's First Zoo*, Philadelphia, 18 (1966), No. 3.

VALERIO, D. A., R. L. MILLER, and J. R. M. INNES: *Macaca mulatta*. New York, London, 1969.

VOGEL, C.: "Ökologie, Lebensweise und Sozialverhalten der Grauen Languren in verschiedenen Biotopen Indiens," in: *Fortschritte der Verhaltensforschung* 17 (1976).

VOGEL, C.: "Zum biologischen Selbstverständnis des Menschen," in: *Naturwiss. Rundschau* 30 (1977).

VOGT, J. L.: "The Social Behavior of a Marmoset (Saguinus fuscicollis) Group III. Spatial Analysis of Social Structure," in: *Folia Primatologica* 29 (1978).

VOSS, H.: *Bibliographie der Menschenaffen*. Jena, 1955.

WAAL, F. B. M. DE: "Schimpansin zieht Stiefkind mit der Flasche auf," in: *Das Tier* 20 (1980), No. 1.

Welt der wilden Tiere. Affen. Time-Life International (Nederland) N. V., 1981.

WENDT, H.: *Auf Noahs Spuren. Die Entdeckung der Tiere*. Hamm, 1956.

WENDT, H.: *Der Affe steht auf*. Reinbek near Hamburg, 1971.

WENDT, H.: *Die Entdeckung der Tiere*. Munich, 1980.

WENDT, H.: *Wir und die Tiere*. Rüschlikon-Zurich, 1954.

WICKLER, W.: *Stammesgeschichte und Ritualisierung*. Munich, 1970.

WILLIAMS, L.: *Der Affe, wie ihn keiner kennt*. Vienna, Munich, Zurich, 1968.

WIRTH, R.: "Auf der Suche nach Madagaskars Geisteraffen," in: *Das Tier* 20 (1980), No. 7.

WOOD, P., L. VACZEK, D. J. JAMBLIN, and J. N. LEONHARD: *Der Weg zum Menschen. Time-Life International* (Nederland) B. V., 1976.

WRANGHAM, R. W.: "An Ecological Model of Female-bonded Primate Groups," in: *Behavior* 75 (1980).

YAMADA, M.: "A Study of Blood Relationship in the Natural Society of the Japanese Macaque," in: *Primates* 4 (1963).

YAMADA, M.: "Five Natural Troops of Japanese Monkeys in Shodoshima Island: Distribution and Social Organization," in: *Primates* 7 (1966).

YERKES, R. M.: *Chimpanzees (A Laboratory Colony)*. London, Oxford, 1943.

YERKES, R. M.: "Gorilla Cenus and Study," in: *Journal of Mammalogy*, 1951.

YERKES, R. M., and A. W. YERKES: *The Great Apes*. New Haven, London, Oxford, 1953.

YLLA: *Animals in India*. Lausanne, 1958.

YOSHIBA, K.: "An Ecological Study of Hanuman Langurs, Presbytis entellus," in: *Primates* 8 (1967).

ZIEMER, L. K.: "Functional Morphology of Forelimb Joints in the Wooly Monkey Lagothrix lagotricha," in: *Contributions to Primatology* 14 (1978).

ZIMMER, D. E.: *Unsere erste Natur. Die biologischen Ursprünge menschlichen Verhaltens*. Munich, 1980.

ZIMMERMANN, E., P. ZIMMERMANN, and A. ZIMMERMANN: "Soziale Kommunikation bei Plumploris (Nycticebus coucang)," in: *Zeitschr. d. Kölner Zoo* 22 (1979), No. 1.

ZIMMERMANN, E., P. ZIMMERMANN, and A. ZIMMERMANN: "Vergleich einiger Kommunikationsformen fünf nonhumaner Primatenarten," in: *Zeitschr. d. Kölner Zoo* 23 (1980), No. 1.

ZISWILER, V.: *Bedrohte und ausgerottete Tiere*. Berlin, Heidelberg, New York, 1965.

ZUCKERMAN, S.: *The Social Life of Monkeys and Apes*. Boston, 1981.

Subject Index

Unitalicized figures refer to page
numbers, figures in italics to numbers
of illustrations.

Conversion factors for measures and weights

1 km	=	0.621370 mile		
1 m	=	3.281 feet	=	39.4 inches
1 cm	=	0.033 foot	=	0.394 inch
1 mm	=	0.0394 inch		
1 l	=	0.2200 gallon (Brit.)		
1 l	=	0.2642 gallon (U.S.)		
1 l	=	1.7598 pints (Brit.), liquid		
1 l	=	2.1134 pints (U.S.), liquid		

The title sheet photos represent:

Page 1: Sumatra Orang Utans *(Pongo pygmaeus abeli)*, Dresden Zoological Gardens; Pages 2/3: Crab-eating Macaques *(Macaca irus)*. "The Sacred Monkey Forest" on Bali; Page 4: Ring-tailed Lemur *(Lemur catta)*, Prague Zoological Gardens